ART NOTEBOOK

EDITION

Discover Biology

ART NOTEBOOK

EDITION

Discover Biology

Michael L. Cain

Hans Damman

Robert A. Lue

Carol Kaesuk Yoon

With contributions from

Richard E. Morel

W. W. NORTON & COMPANY
NEW YORK • LONDON

Third Edition

ISBN-10: 0-393-92846-2
ISBN-13: 978-0-393-92846-4

W. W. Norton & Company, Inc., 500 Fifth Avenue, New York, NY 10110
  www.wwnorton.com

W. W. Norton & Company Ltd., Castle House, 75/76 Wells Street, London W1T 3QT

2 3 4 5 6 7 8 9 0

Contents

Unit 1—The Diversity of Life

Chapter 1: The Nature of Science and the Characteristics of Life 1
Chapter 2: Organizing the Diversity of Life 6
Chapter 3: Major Groups of Living Organisms 11

INTERLUDE A —Biodiversity and People 20

Unit 2—Cells: The Basic Units of Life

Chapter 4: Chemical Building Blocks 22
Chapter 5: Cell Structure and Compartments 34
Chapter 6: Cell Membranes, Transport, and Communication 45
Chapter 7: Energy and Enzymes 53
Chapter 8: Photosynthesis and Respiration 60
Chapter 9: Cell Division 68

INTERLUDE B —Cancer: Cell Division Out of Control 74

Unit 3—Genetics

Chapter 10: Patterns of Inheritance 79
Chapter 11: Chromosomes and Human Genetics 85
Chapter 12: DNA 94
Chapter 13: From Gene to Protein 101
Chapter 14: Control of Gene Expression 108
Chapter 15: DNA Technology 114

INTERLUDE C —Harnessing the Human Genome 122

Unit 4—Evolution

Chapter 16: How Evolution Works 127
Chapter 17: Evolution of Populations 134
Chapter 18: Adaptation and Speciation 140
Chapter 19: The Evolutionary History of Life 145

INTERLUDE D —Humans and Evolution 156

Unit 5—Animal Form and Function

Chapter 20: Maintaining the Internal Environment 161
Chapter 21: Nutrition and Digestion 167
Chapter 22: Gas Exchange 176
Chapter 23: Blood and Circulation 182
Chapter 24: Animal Hormones 189
Chapter 25: The Nervous System 195
Chapter 26: Sensing the Environment 202
Chapter 27: Muscles, Skeletons, and Movement 209
Chapter 28: Defense Against Disease 218
Chapter 29: Reproduction and Development 224
Chapter 30: Animal Behavior 234

INTERLUDE E —Smoking: Beyond Lung Cancer 237

Unit 6—Plant Form and Function

Chapter 31: Plant Structure, Nutrition, and Transport 239
Chapter 32: Plant Growth and Reproduction 247

INTERLUDE F —Feeding a Hungry Planet 255

Unit 7—Interactions with the Environment

Chapter 33: The Biosphere 258
Chapter 34: Growth of Populations 263
Chapter 35: Interactions among Organisms 270
Chapter 36: Communities of Organisms 274
Chapter 37: Ecosystems 278
Chapter 38: Global Change 288

INTERLUDE G —Building a Sustainable Society 293

CHAPTER 1

The Nature of Science and the Characteristics of Life

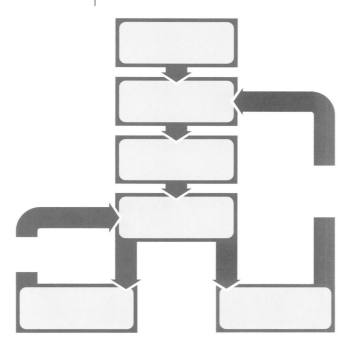

Figure 1.2 The Scientific Method

Table 1.1

The Shared Characteristics of Life
All living organisms
• are built of cells
• reproduce themselves using the hereditary material DNA
• develop
• capture energy from their environment
• sense their environment and respond to it
• show a high level of organization
• evolve

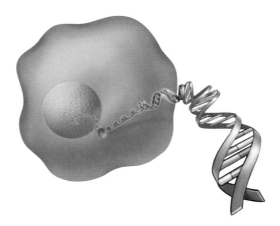

Figure 1.5 The DNA Molecule: A Blueprint for Life

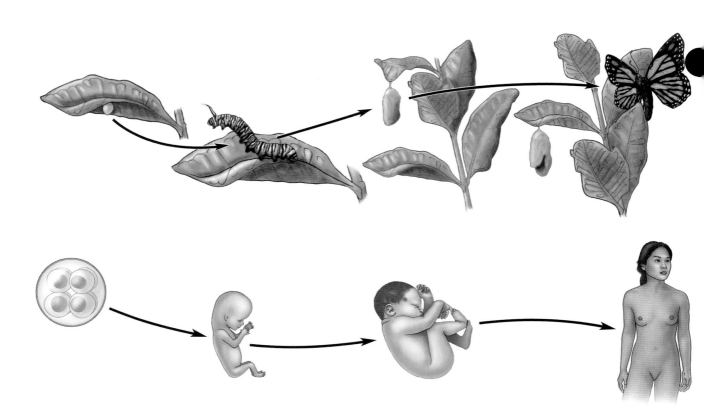

Figure 1.6 Growing Up

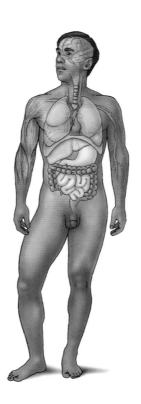

Figure 1.9 Staying Organized

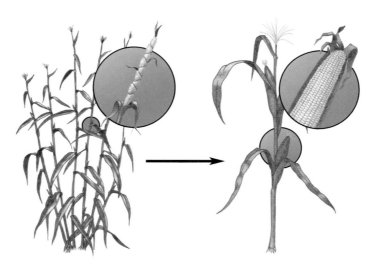

Figure 1.10 Living Organisms Evolve

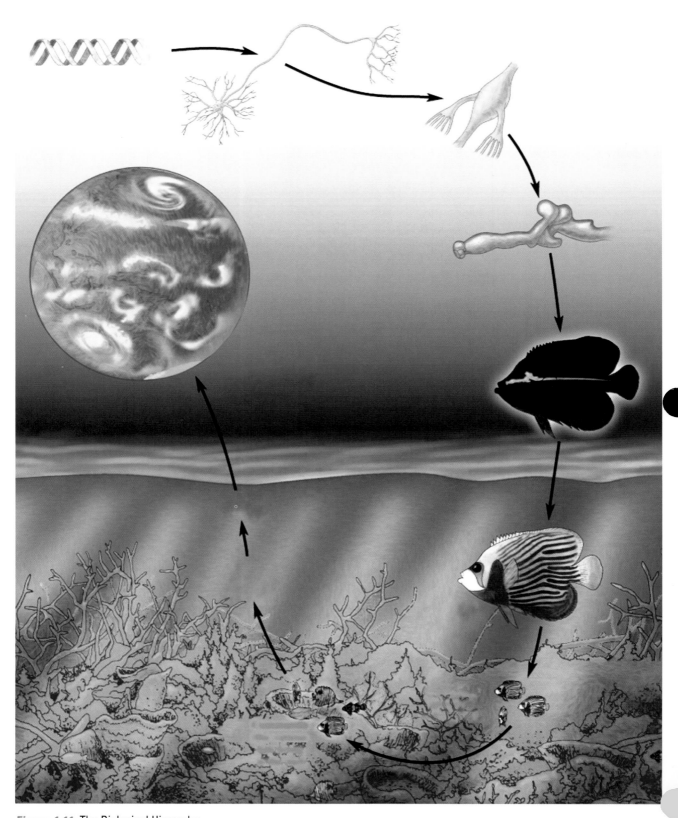

Figure 1.11 The Biological Hierarchy

Figure 1.13 A Desert Food Web

CHAPTER 2

Organizing the Diversity of Life

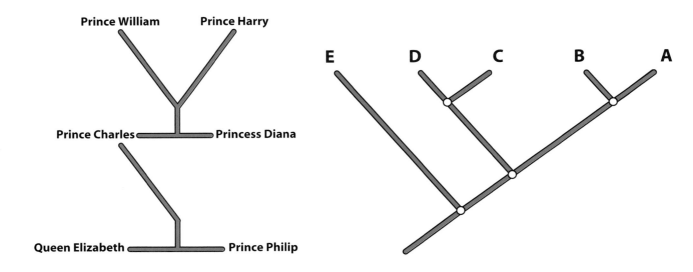

Figure 2.1 Family Trees versus Evolutionary Trees

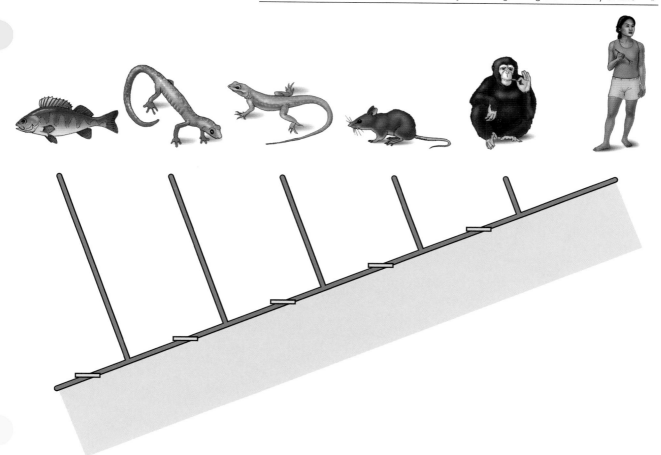

Figure 2.2 Shared Derived Features Define Evolutionary Relationships

Figure 2.4 The Close Relationship Between Birds and Dinosaurs

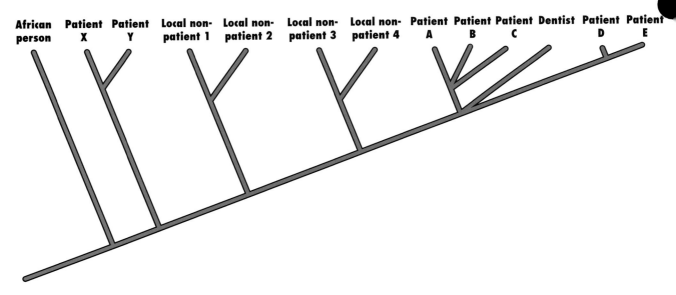

Science Toolkit An Evolutionary Tree Solves the Whodunit

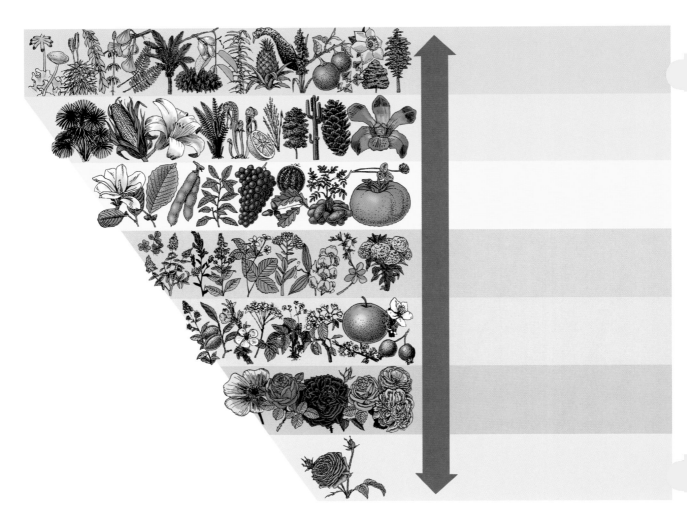

Figure 2.6 The Linnaean Hierarchy

Figure 2.7 Kingdoms and Domains

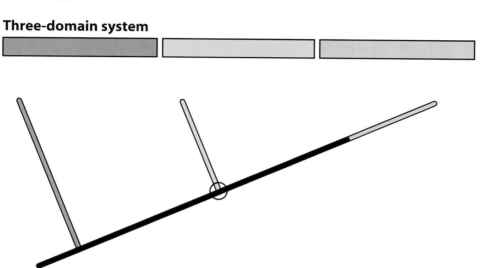

Figure 2.8 Evolutionary Tree of Domains

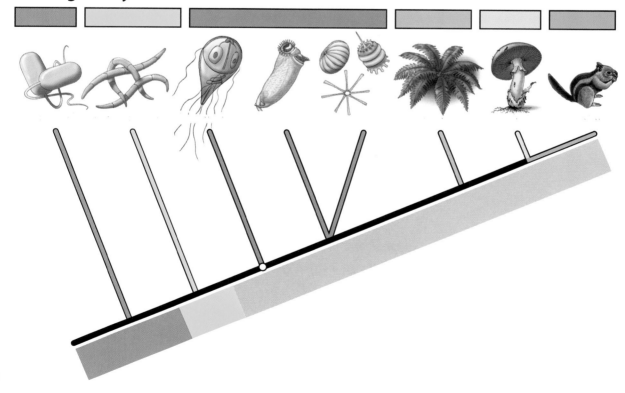

Figure 2.9 The Tree of All Life

Figure 2.11 The Primate Evolutionary Tree

Representatives of Eukarya

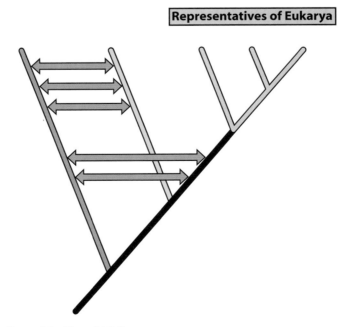

Figure 2.12 A Tangled Web at the Base of the Tree of Life?

CHAPTER 3

Major Groups of Living Organisms

Domains

Kingdoms of the Linnaean hierarchy

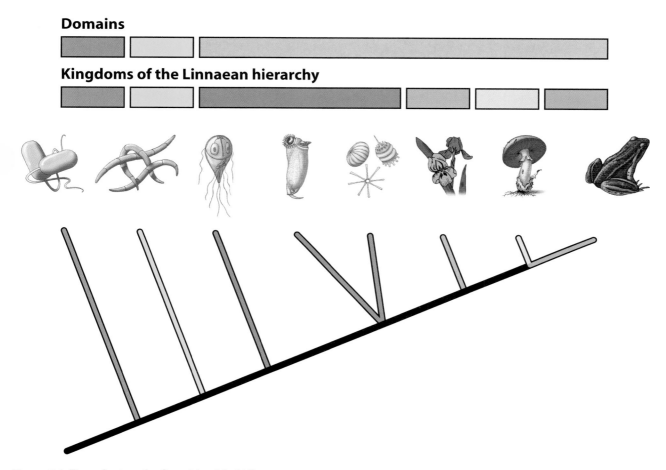

Figure 3.1 Three Systems for Organizing All of Life

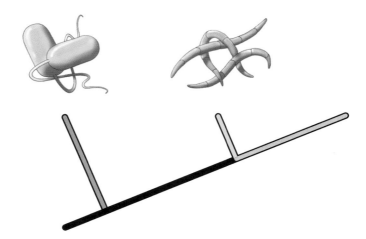

Figure 3.2 The Prokaryotes: Bacteria and Archaea

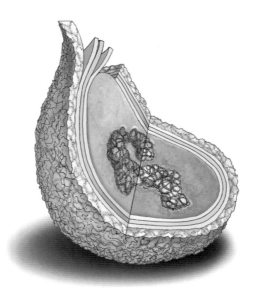

Figure 3.3 The Basic Structure of the Prokaryotic Cell

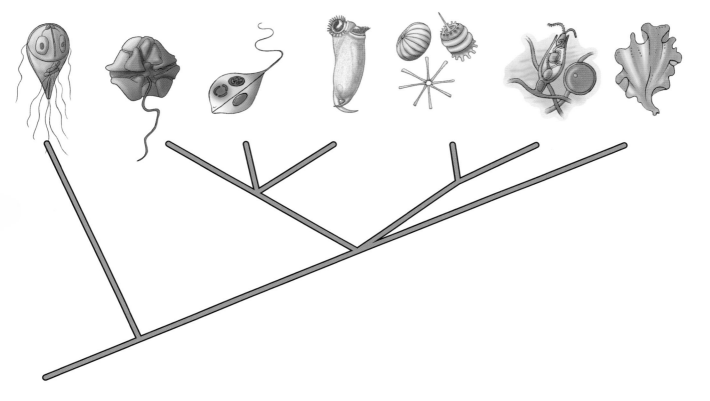

Figure 3.7 The Protista

Figure 3.8 The Plantae

Figure 3.9 The Basic Structures of a Plant

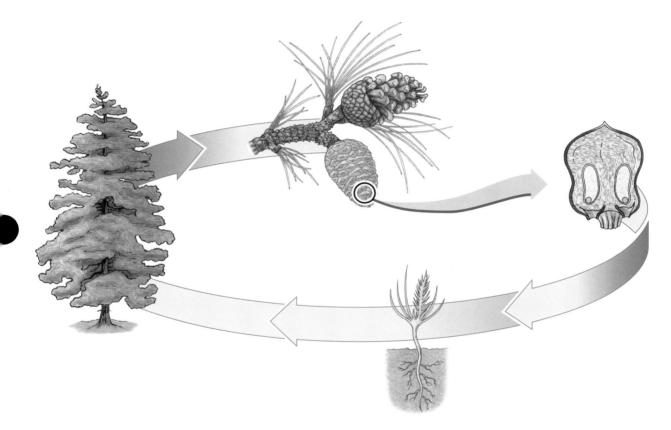

Figure 3.10 The Seed

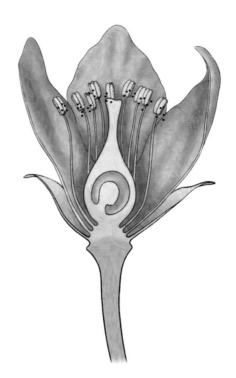

Figure 3.11 The Flower

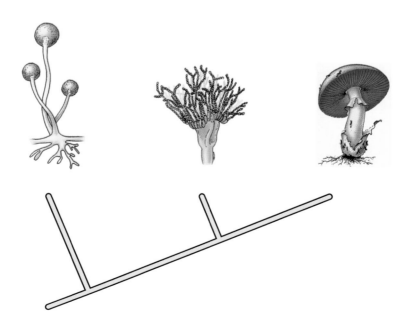

Figure 3.13 The Fungi

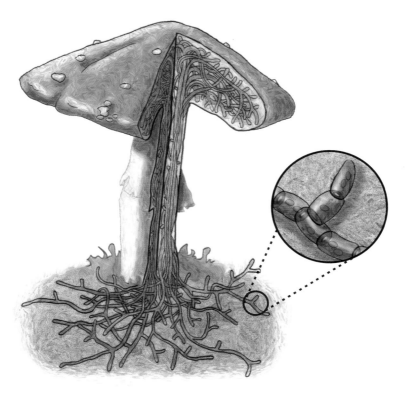

Figure 3.14 The Basic Structures of a Fungus

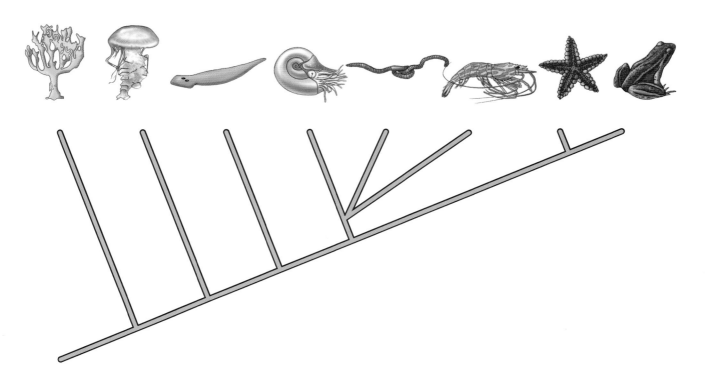

Figure 3.18 The Animalia

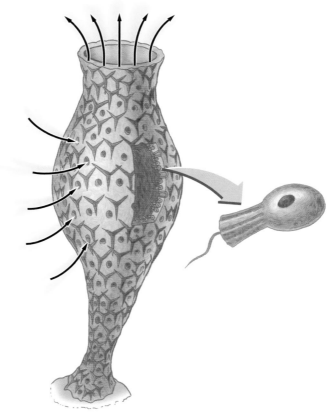

Figure 3.19 Sponges Have Specialized Cells but Lack Tissues

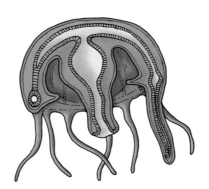

Figure 3.20 Jellyfish Have True Tissue Layers

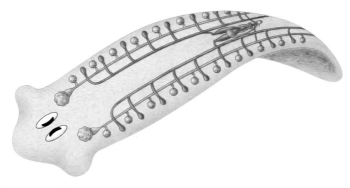

Figure 3.21 Flatworms Evolved Organs and Organ Systems

Figure 3.22 Variations on a Theme

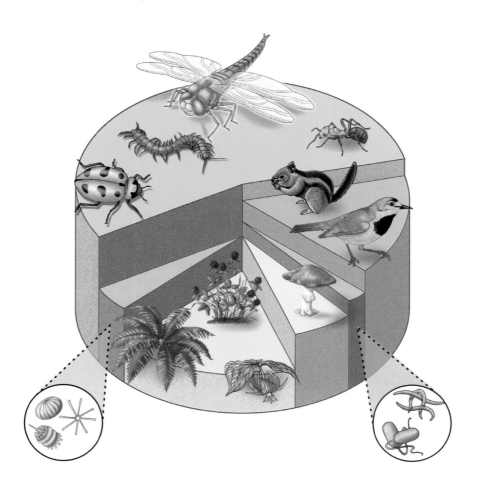

Figure A.3 A Piece of the Pie

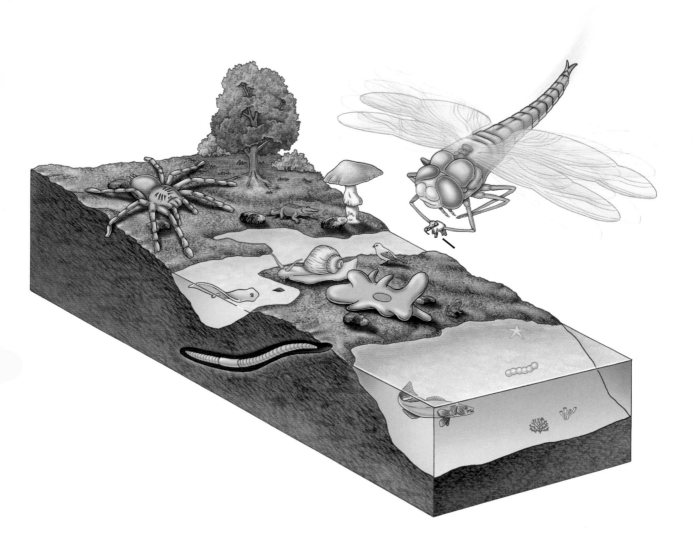

Figure A.4 A Species-Scape

CHAPTER 4 | Chemical Building Blocks

Figure 4.1 Atoms

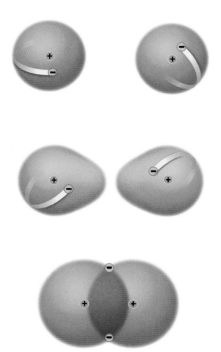

Atom	Symbol	Number of possible bonds	Sample compounds
Hydrogen			
Oxygen			
Sulfur			
Nitrogen			
Carbon			
Phosphorus			

Figure 4.3 Covalent Bonds

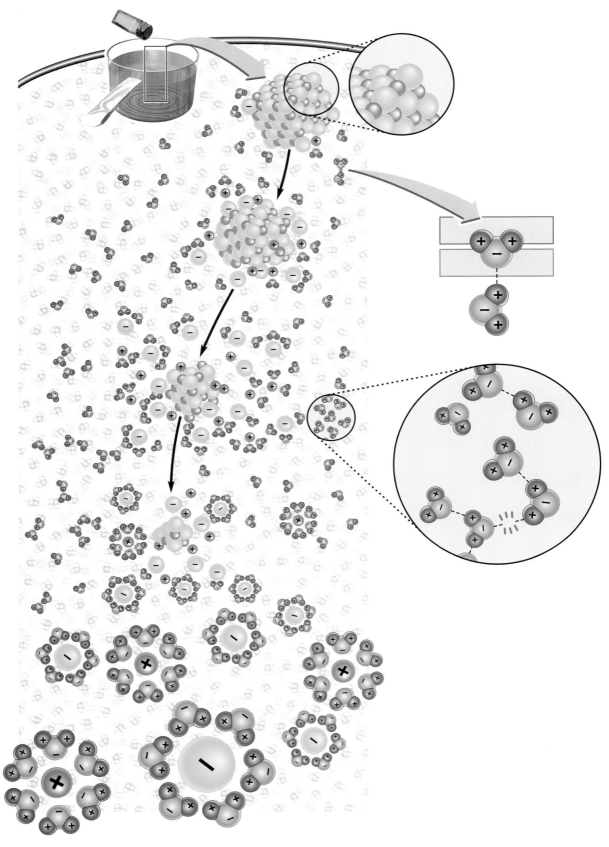

Figure 4.4 Hydrogen Bonds Determine the Properties of Water and How Other Compounds Interact with Water

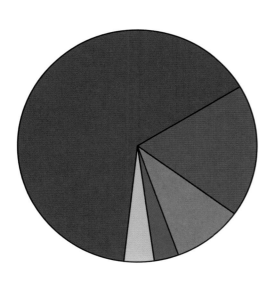

Biology Matters Elements 'R Us

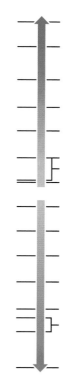

Figure 4.5 The pH Scale

Table 4.1

Some Important Functional Groups Found in Organic Molecules

Functional group	Formula	Ball-and-stick model
Amino group	—NH₂	Bond to carbon atom
Carboxyl group	—COOH	
Hydroxyl group	—OH	
Phosphate group	—PO₄	

	Name	Space-filling model	Ball-and-stick model	Structural formula
Monomer				
Polymers				

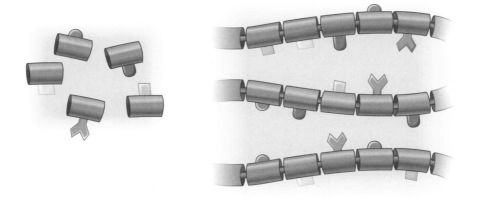

Figure 4.6 Assembling Complex Structures from Smaller Components

Figure 4.7 The Structure of Carbohydrates

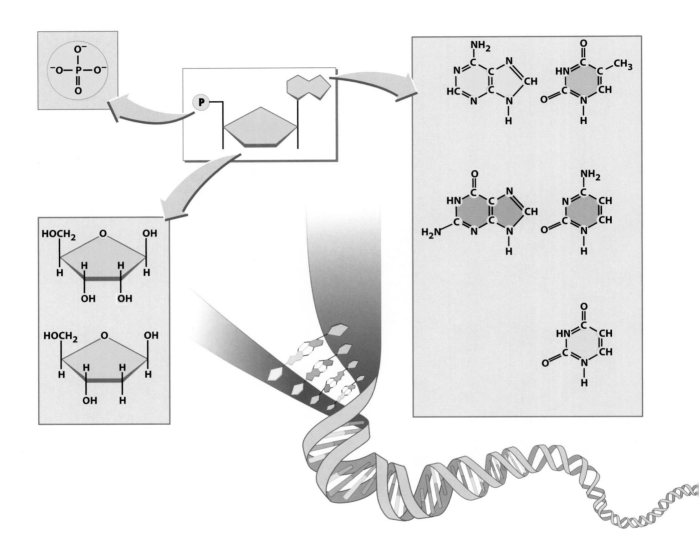

Figure 4.8 Nucleotide Components

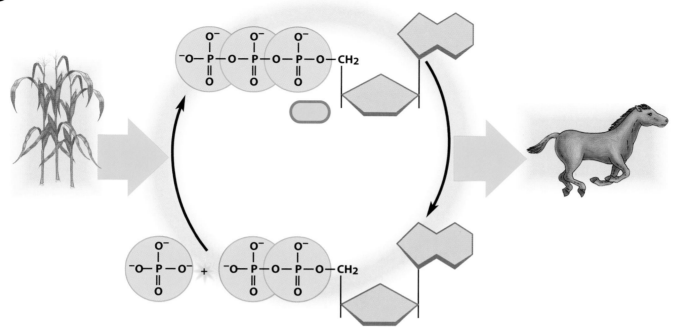

Figure 4.9 Production of ATP

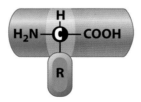

Figure 4.10 The Structure of Amino Acids

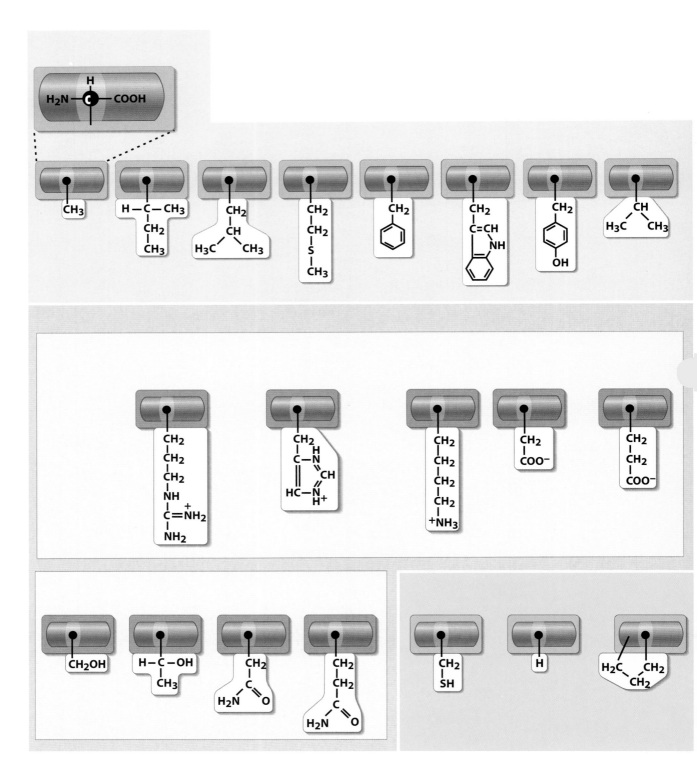

Figure 4.11 The Diversity of Amino Acids

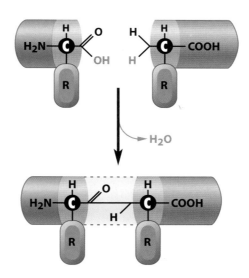

Figure 4.12 Formation of a Peptide Bond

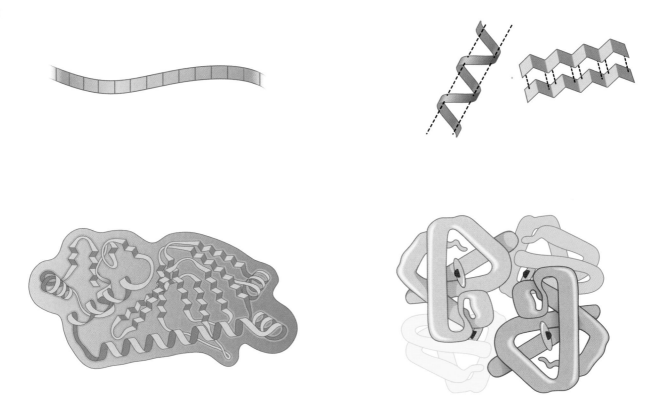

Figure 4.13 The Four Levels of Protein Structure

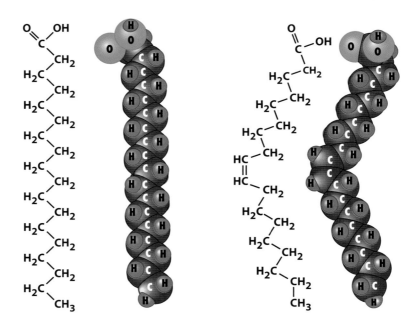

Figure 4.14 Saturated and Unsaturated Fatty Acids

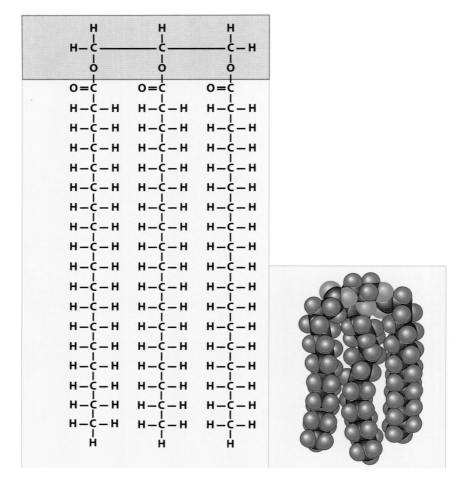

Figure 4.15 Fats

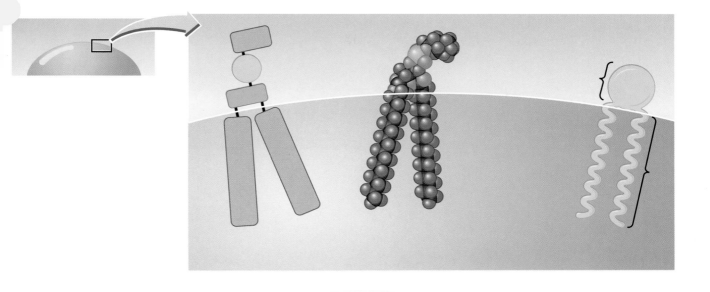

Figure 4.16 Phospholipids Form Membranes

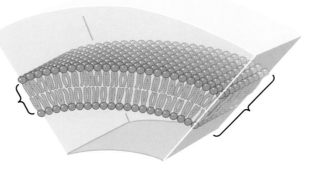

Figure 4.17 Steroids

CHAPTER 5

Cell Structure and Compartments

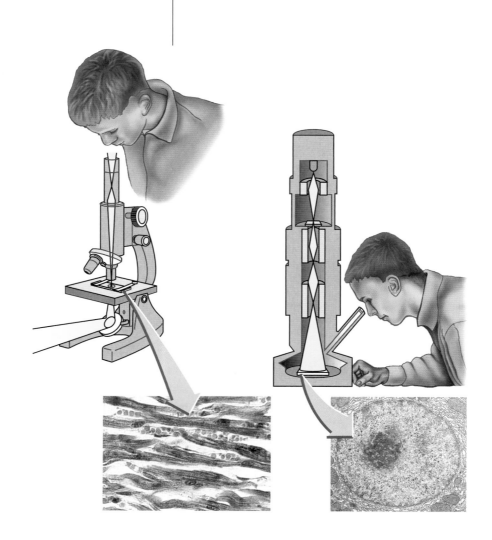

Science Toolkit Exploring Cells under the Microscope

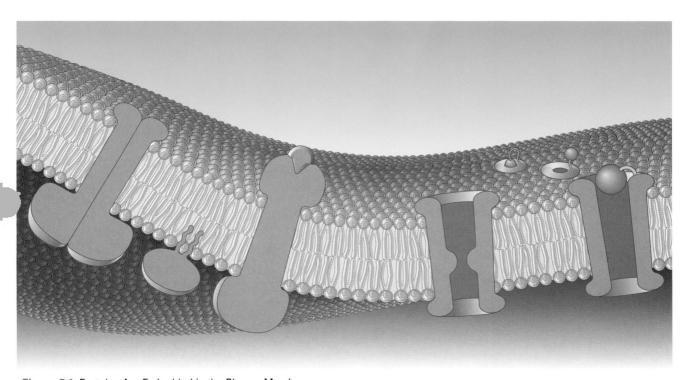

Figure 5.1 Proteins Are Embedded in the Plasma Membrane

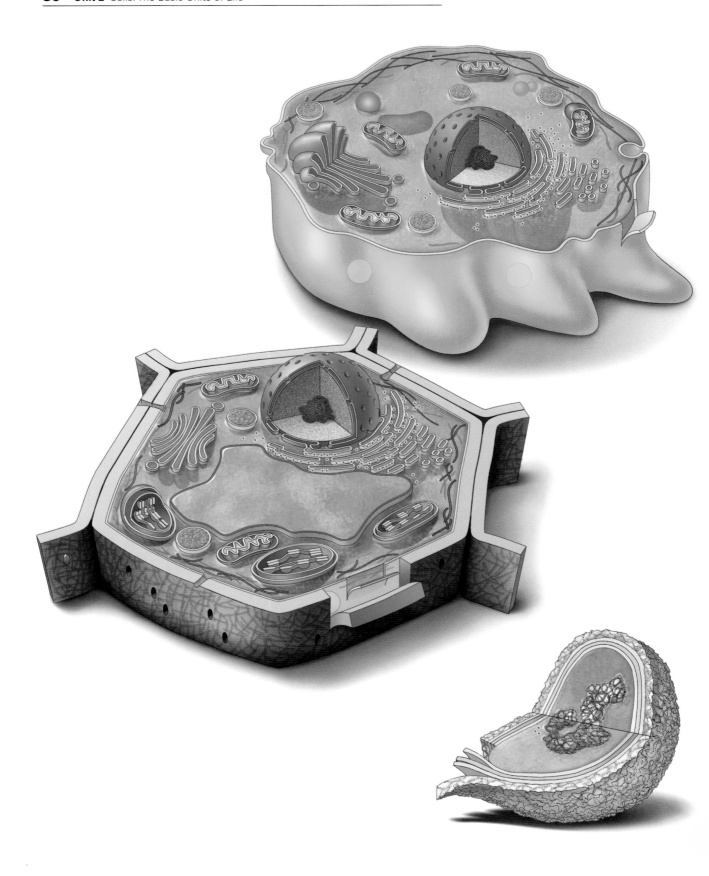

Figure 5.2 Prokaryotic and Eukaryotic Cells Compared

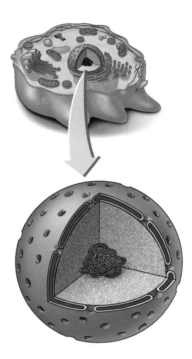

Figure 5.3 The Nucleus

Figure 5.4 The Endoplasmic Reticulum

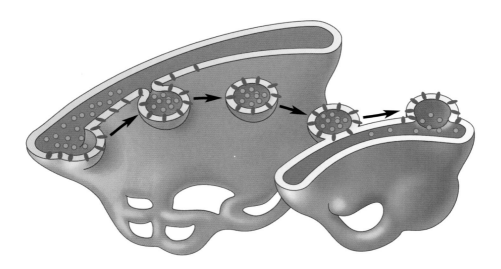

Figure 5.5 How Vesicles Move Proteins and Lipids from One Compartment to Another

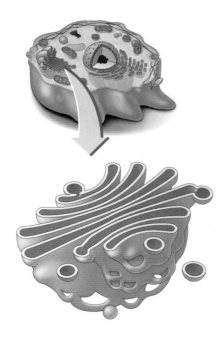

Figure 5.6 The Golgi Apparatus

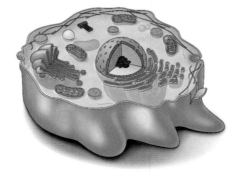

Figure 5.7 Lysosomes in an Animal Cell

Figure 5.8 Vacuoles in a Plant Cell

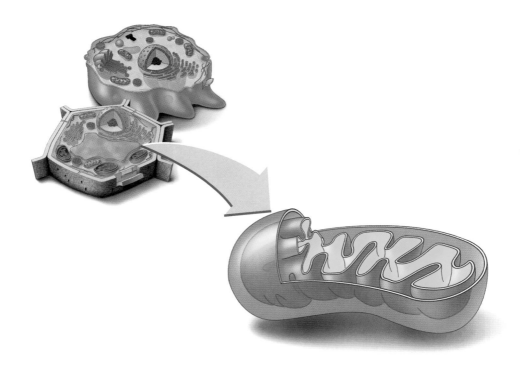

Figure 5.9 Energy-Transforming Organelles: Mitochondria

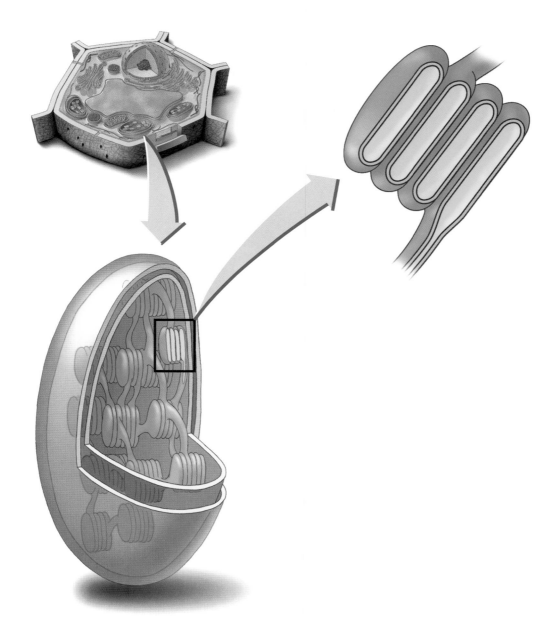

Figure 5.10 Energy-Transforming Organelles: Chloroplasts

Figure 5.11 Kinds of Filaments in the Cytoskeleton

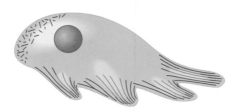

Figure 5.12 Microfilaments Allow Cell Movement

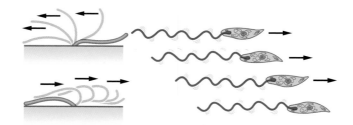

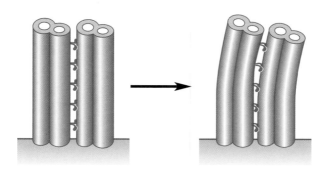

Figure 5.13 Eukaryotic Cilia and Flagella

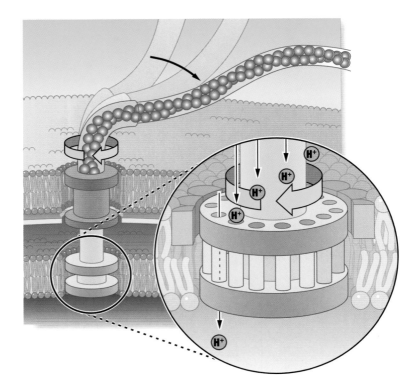

Figure 5.14 Bacterial Flagella

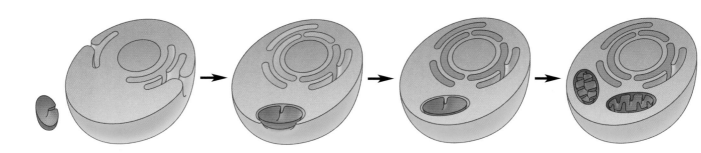

Figure 5.15 How Primitive Eukaryotes May Have Acquired Organelles

CHAPTER 6

Cell Membranes, Transport, and Communication

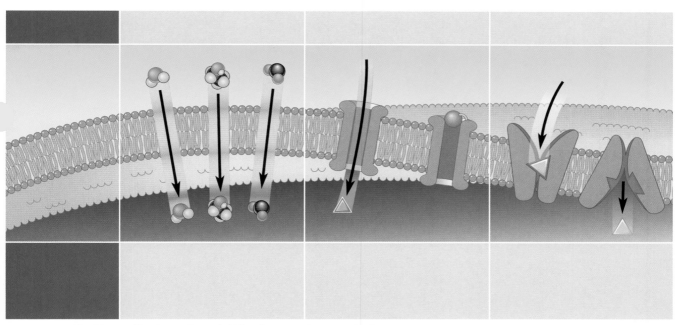

Figure 6.1 The Plasma Membrane Controls What Enters and Leaves the Cell

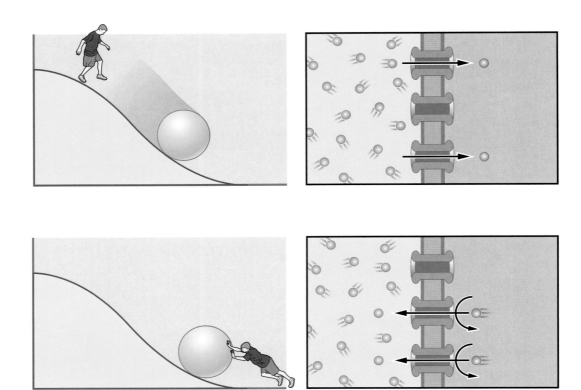

Figure 6.2 Active versus Passive Movement of Molecules

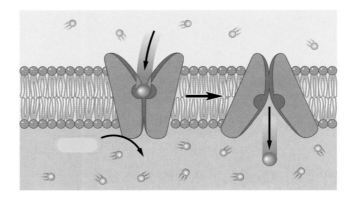

Figure 6.3 Active Carrier Proteins

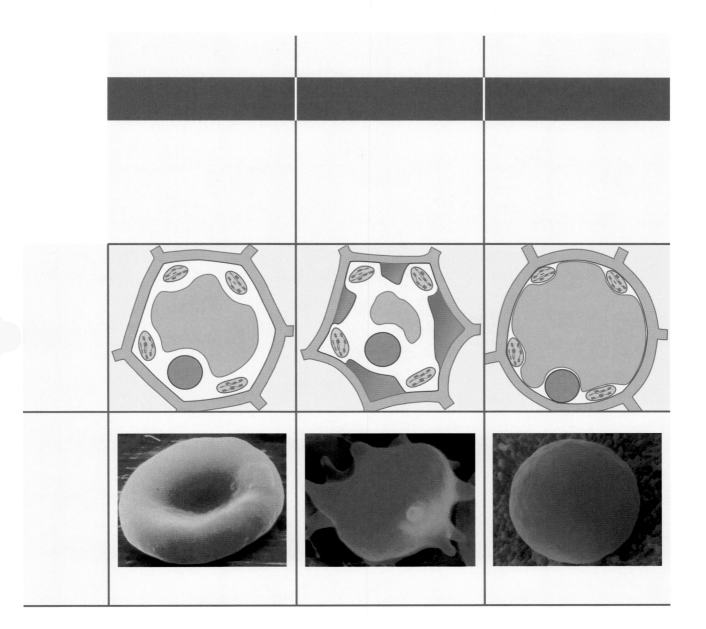

Figure 6.4 Water Moves Into and Out of Cells by Osmosis

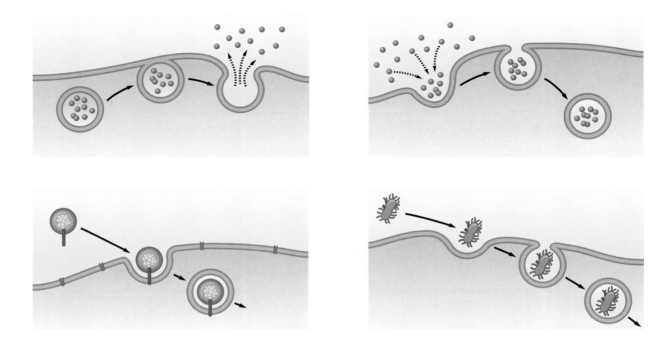

Figure 6.5 Cells Are Importers and Exporters

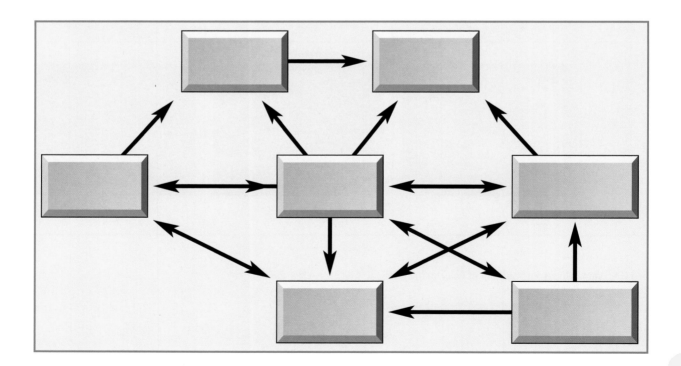

Figure 6.6 The Cellular Functions Required for Life

Figure 6.7 Cells in Multicellular Organisms Are Interconnected in Various Ways

Figure 6.8 Receptors for Signaling Molecules

Table 6.1

Examples of Signaling Molecules

Type of molecule	Name of molecule	Site(s) of synthesis	Function(s)
ANIMALS			
Amino acid derivatives	Adrenaline	Adrenal glands	Promotes release of stored fuels Promotes increased heart rate
	Thyroxine	Thyroid gland	Promotes increased metabolic rate
Choline derivative	Acetylcholine	Nerve cells	Assists signal transmission from nerves to muscles
Gas	Nitric oxide	Endothelial cells in blood vessel walls	Promotes relaxation of blood vessel walls
		Nerve cells	
Proteins	Insulin	Beta cells of the pancreas	Promotes the uptake of glucose by cells
	Nerve growth factor (NGF)	Tissues richly supplied with nerves	Promotes nerve growth and survival
	Platelet-derived growth factor (PDGF)	Many cell types	Promotes cell division
Steroids	Progesterone	Ovaries	Prepares the uterus for implantation Promotes mammary gland development
	Testosterone	Testes	Promotes the development of secondary male sex characteristics
PLANTS			
Acetic acid derivative	Auxin	Most plant cells	Promotes root formation Promotes stem elongation

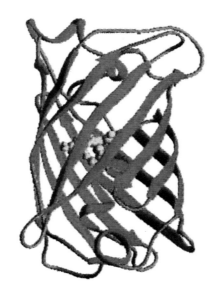

Science Toolkit, Figure A Molecular Model of the Lantern Protein

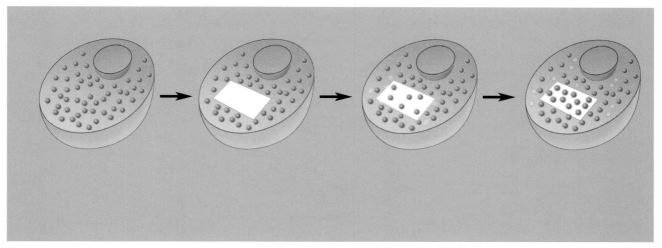

Science Toolkit, Figure B Tracking the Movement of Proteins Tagged with the Lantern Protein

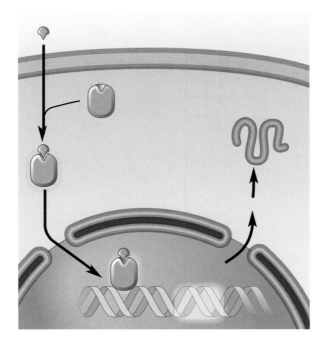

Figure 6.9 A Cell's Response to a Steroid Hormone

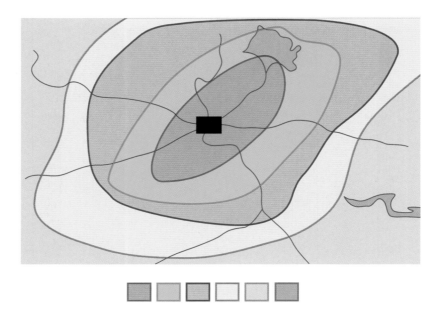

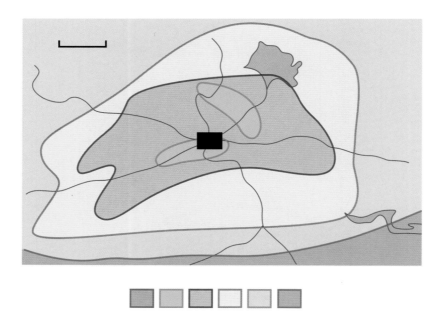

Figure 6.10 The Effect of Sulfur Dioxide on Lichen Diversity

CHAPTER 7 Energy and Enzymes

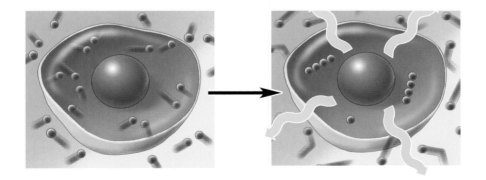

Figure 7.1 The Second Law of Thermodynamics

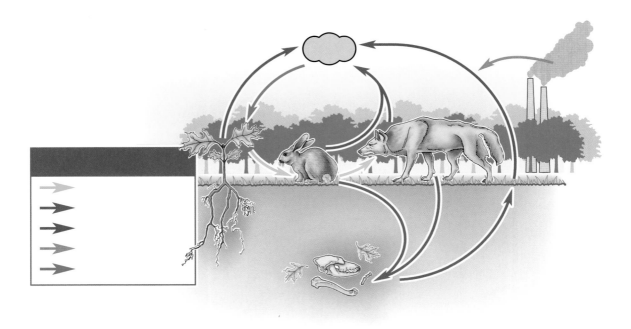

Figure 7.2 Carbon Cycling

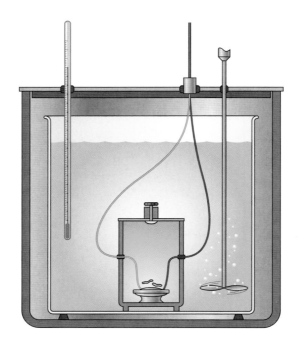

Science Toolkit Inside a Bomb Calorimeter

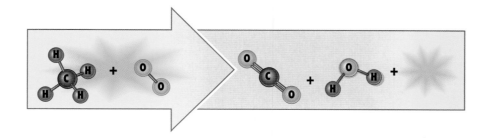

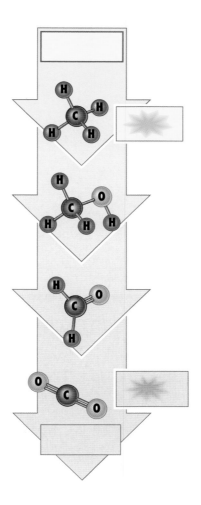

Figure 7.3 Oxidation of Methane

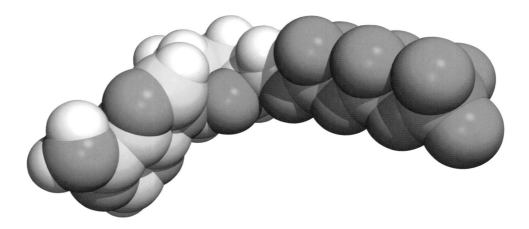

Figure 7.4 ATP Molecule

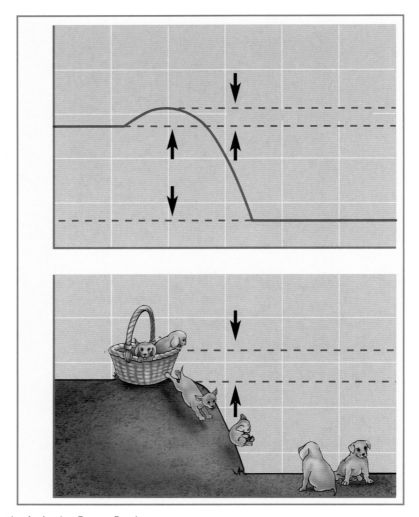

Figure 7.5 Getting Over the Activation Energy Barrier

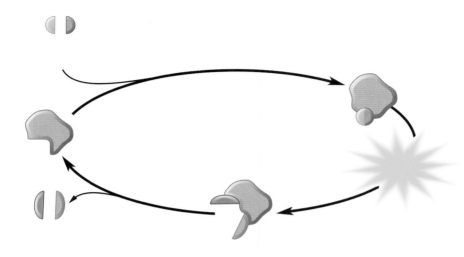

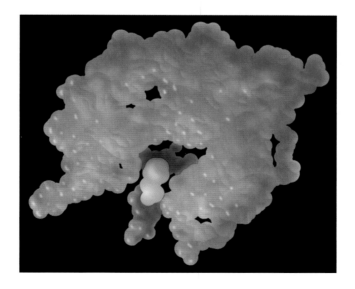

Figure 7.6 Enzymes as Molecular Matchmakers

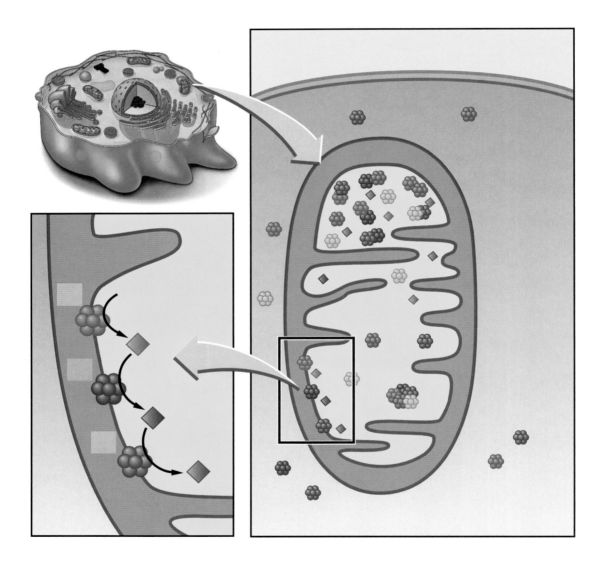

Figure 7.7 Grouping of Enzymes in the Cell

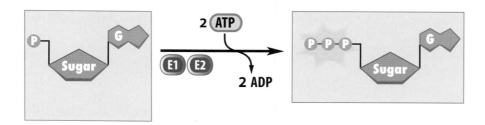

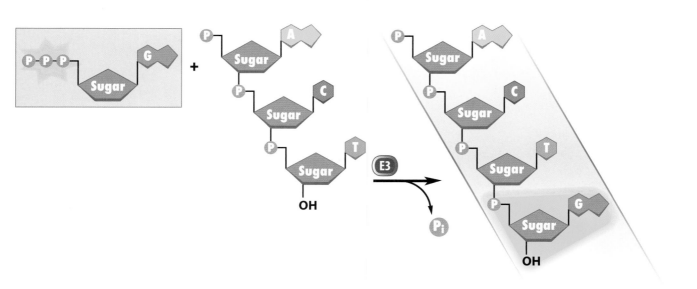

Figure 7.8 Biosynthesis of DNA

CHAPTER 8

Photosynthesis and Respiration

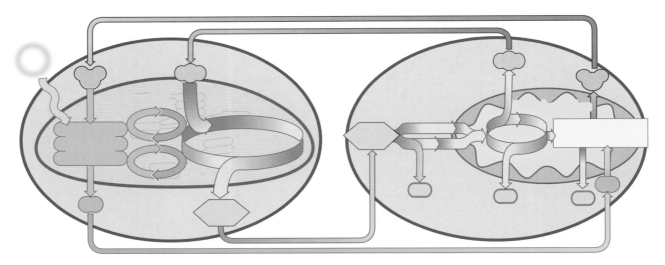

Figure 8.1 The Exchange of Molecules Between Chloroplasts and Mitochondria Produces Energy Carriers

Figure 8.2 The Flow of Energy Between Organisms and Their Environment

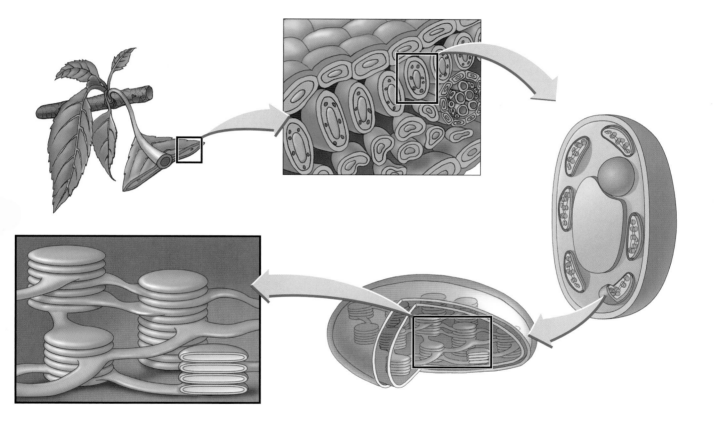

Figure 8.3 Chloroplasts in a Leaf Cell

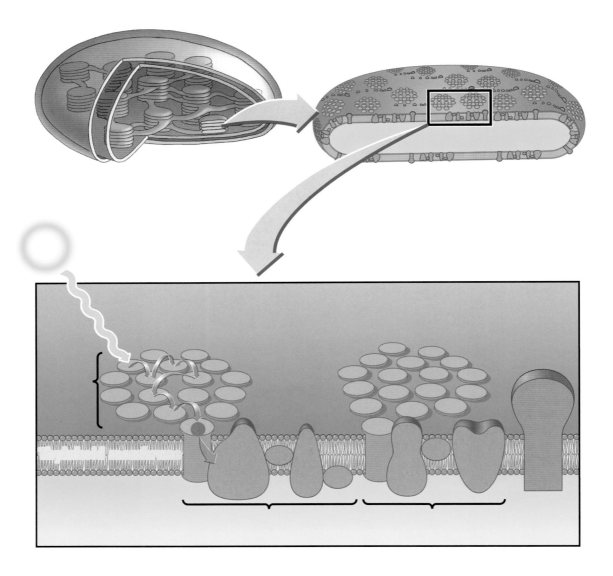

Figure 8.4 The Arrangement of Photosystems in the Thylakoid Membrane

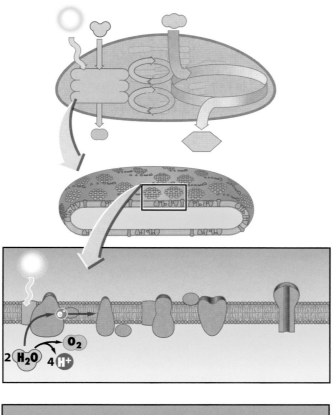

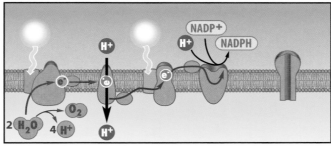

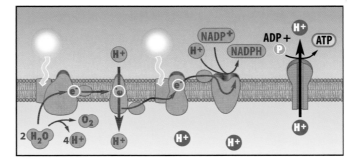

Figure 8.5 Production of Energy Carriers by the Light Reactions

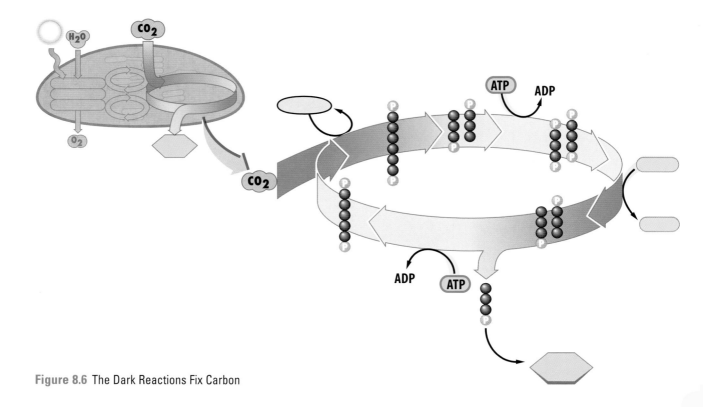

Figure 8.6 The Dark Reactions Fix Carbon

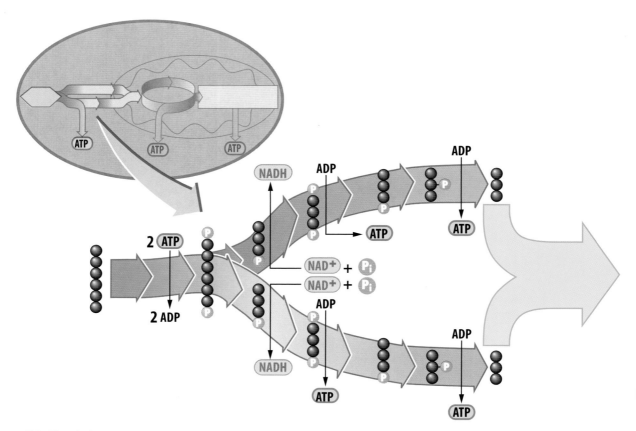

Figure 8.8 Glycolysis

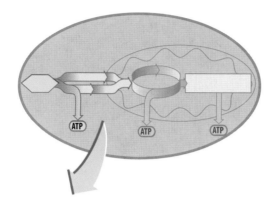

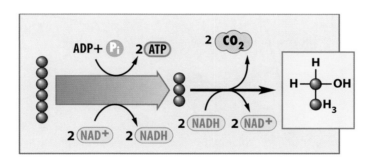

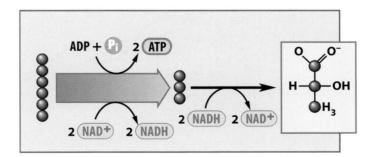

Figure 8.9 Fermentation Has a Variety of Uses

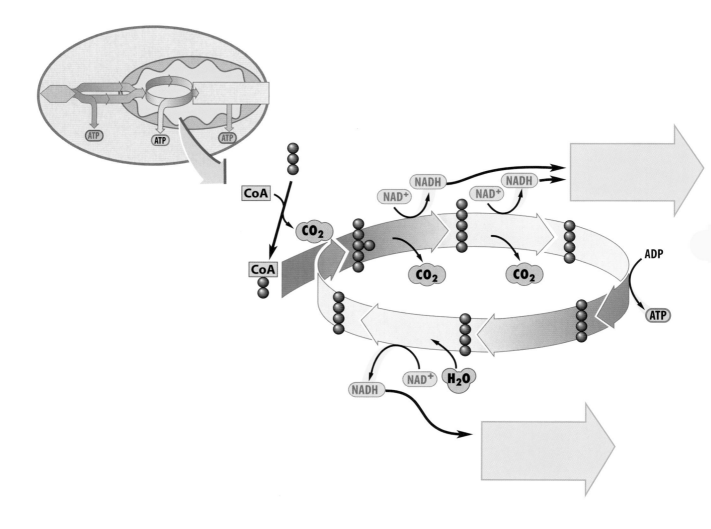

Figure 8.10 The Citric Acid Cycle

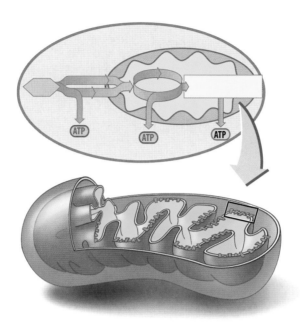

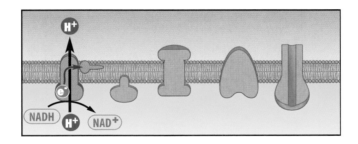

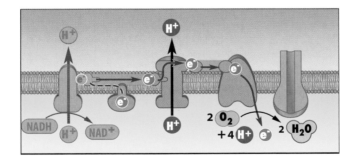

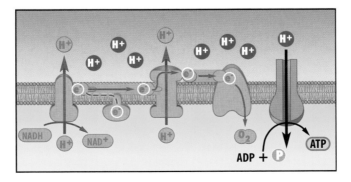

Figure 8.11 Oxidative Phosphorylation

CHAPTER 9

Cell Division

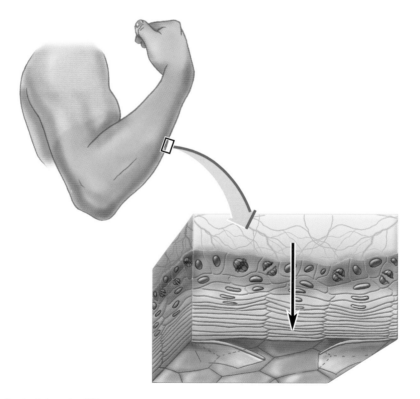

Figure 9.1 Cell Division Replenishes the Skin

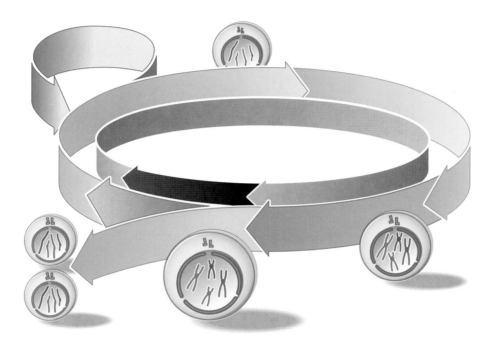

Figure 9.2 The Cell Cycle

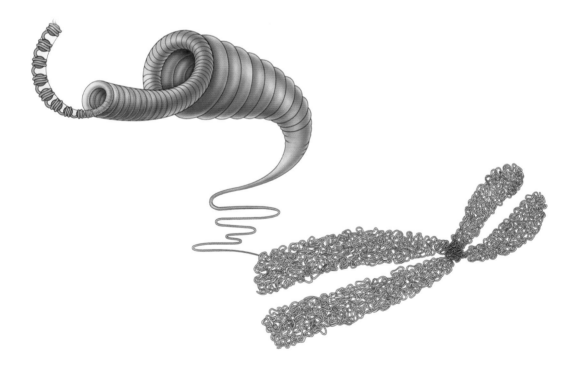

Figure 9.3 The Packing of DNA into a Chromosome

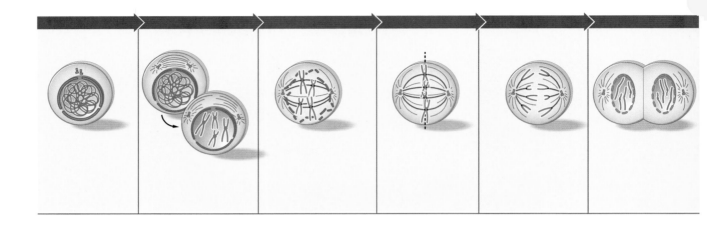

Figure 9.5 The Stages of Cell Division

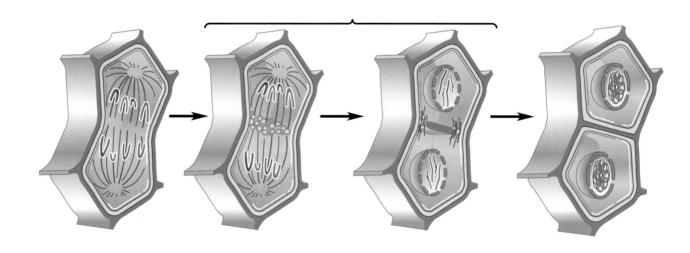

Figure 9.6 Cell Division in Plants

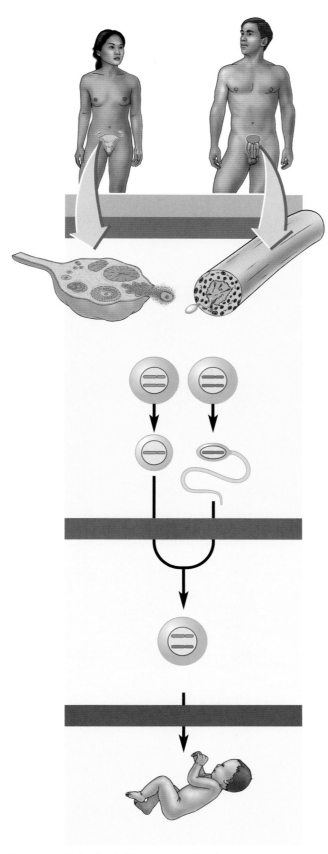

Figure 9.7 Sexual Reproduction Requires a Reduction in Chromosome Number

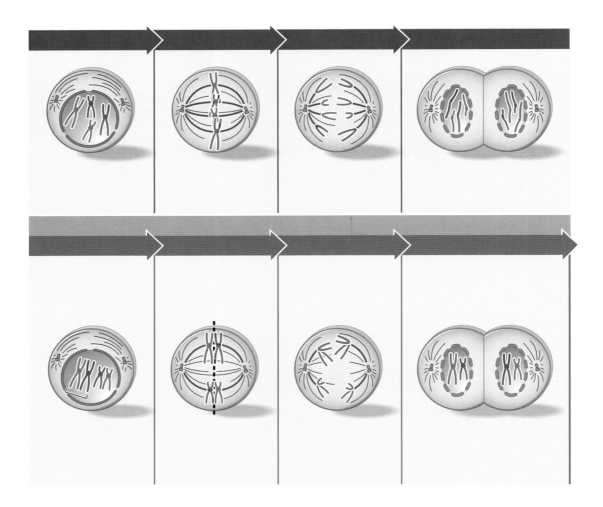

Figure 9.8 Similarities and Differences Between Meiosis and Mitosis

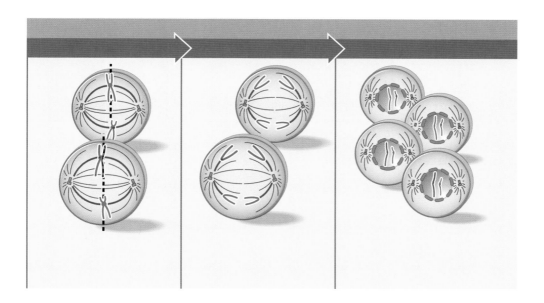

Cancer: Cell Division Out of Control

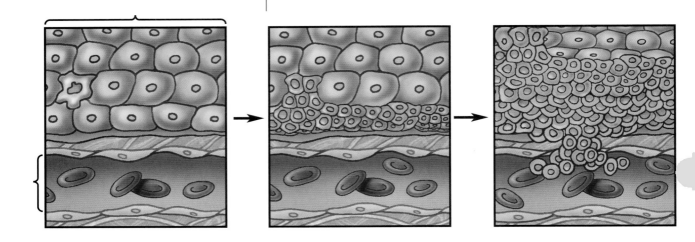

Figure B.2 Cancers Start with a Single Cell That Loses Control

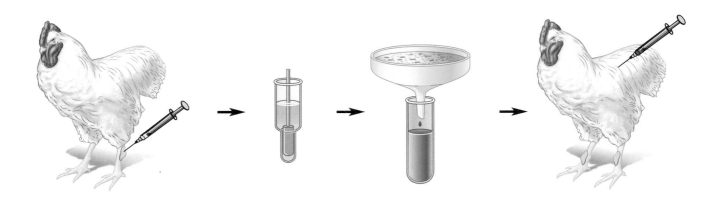

Figure B.3 The Rous Sarcoma Virus Causes Cancer in Chickens

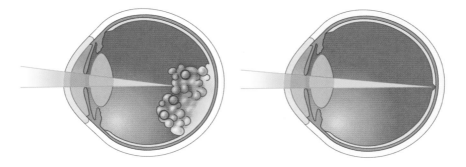

Figure B.5 Retinoblastoma

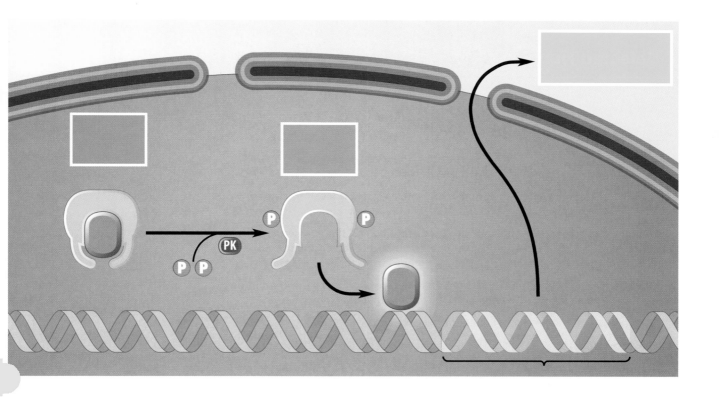

Figure B.6 How the Rb Protein Inhibits Cell Division

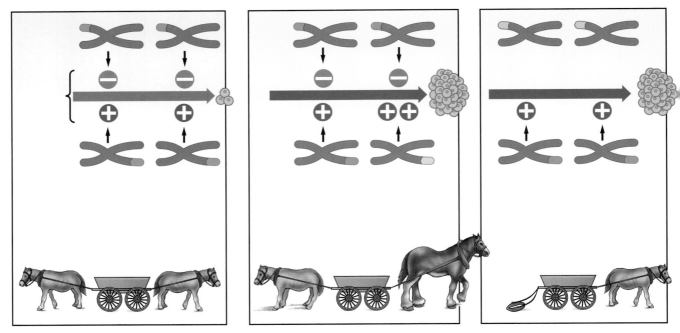

Figure B.7 The Control of Cell Division by Proto-oncogenes and Tumor Suppressor Genes

Table B.1

Selected Human Cancers in the United States

Type of cancer	Observation	Estimated new cases in 2005	Estimated deaths in 2005
Breast cancer	The second leading cause of cancer deaths in women	211,200	40,400
Colon and rectal cancer	The number of new cases is leveling off as a result of early detection and polyp removal	145,300	56,300
Leukemia	Often thought of as a childhood disease, this cancer of white blood cells affects more than 10 times as many adults as children every year	38,800	22,600
Lung cancer	Accounts for 28 percent of all cancer deaths and kills more women than breast cancer does	172,600	163,500
Ovarian cancer	Accounts for 3 percent of all cancers in women	22,200	16,200
Prostate cancer	The second leading cause of cancer deaths in men	232,100	30,400
Malignant melanoma	The most serious and rapidly increasing form of skin cancer in the United States	59,600	10,600

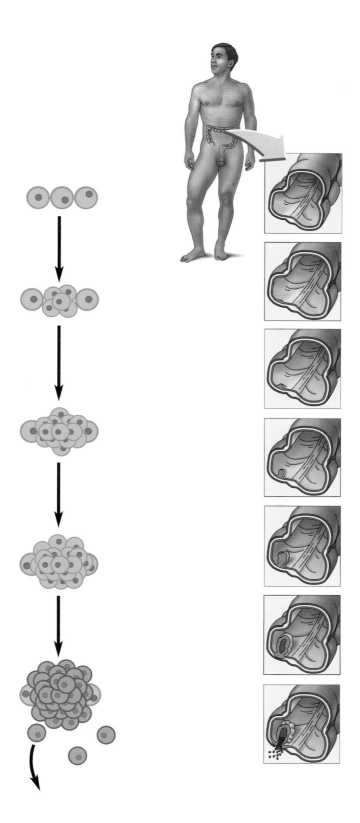

Figure B.8 Colon Cancer Is a Multi-step Process

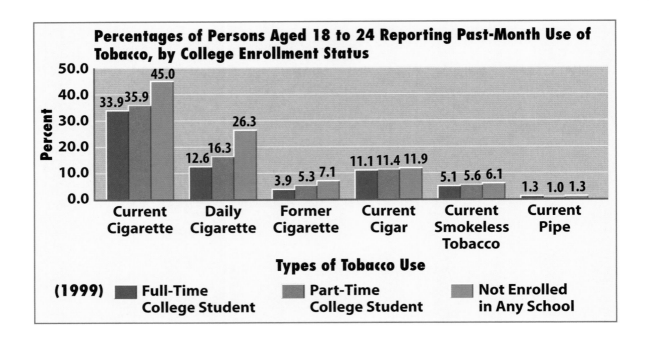

Percentages of Persons Aged 18 to 24 Reporting Past-Month Use of Tobacco, by College Enrollment Status

Biology Matters The Truth About Cigarettes in School

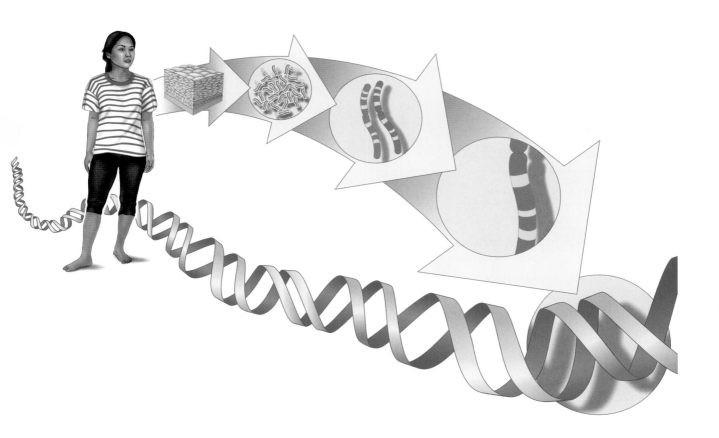

Figure 10.2 We Have the Same Genes in All Our Cells

Figure 10.3 Genotype and Phenotype

Table 10.1

Basic Terms in Genetics

Term	Definition
Allele	One of two or more alternative versions of a gene.
Dominant allele	An allele that determines the phenotype of an organism even when paired with a different (recessive) allele.
F_1 generation	The first generation of offspring in a genetic cross.
F_2 generation	The second generation of offspring in a genetic cross.
Gene	An individual unit of genetic information for a specific trait. Genes are located on chromosomes and are the basic unit of inheritance.
Genetic cross	A controlled mating experiment, usually performed to examine the inheritance of a particular trait.
Genotype	The genetic makeup of an organism.
Heterozygote	An individual that carries one copy of each of two different alleles (for example, an Aa individual or a $C^W C^R$ individual).
Homozygote	An individual that carries two copies of the same allele (for example, an AA, aa, or $C^W C^W$ individual).
P generation	The parent generation of a genetic cross.
Phenotype	The observable characteristics of an organism.
Recessive allele	An allele that does not have a phenotypic effect when paired with a dominant allele.
Trait	A feature of an organism, such as height, flower color, or the chemical structure of a protein.

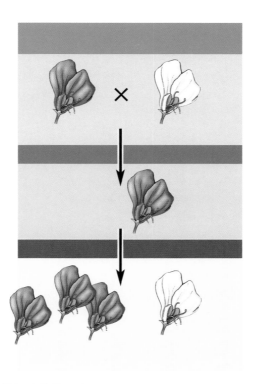

Figure 10.5 Three Generations in One of Mendel's Experiments

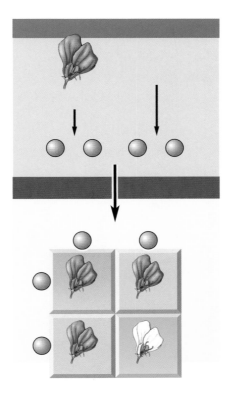

Figure 10.6 The Punnett Square Method

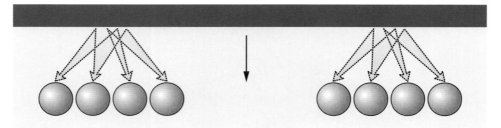

Figure 10.7 Independent Assortment of Alleles

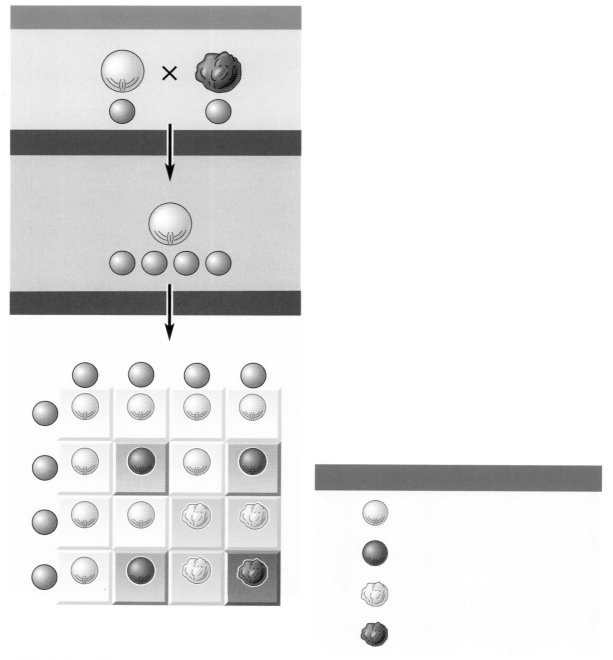

Figure 10.8 Are Genes Inherited Independently?

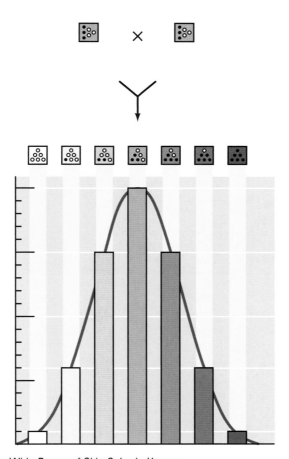

Figure 10.13 Three Genes Produce a Wide Range of Skin Color in Humans

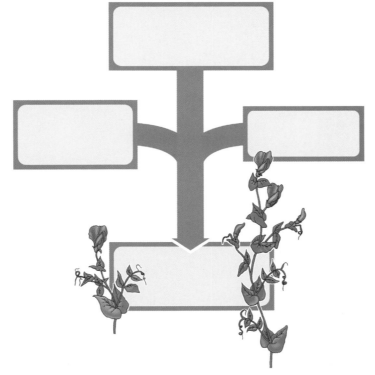

Figure 10.14 From Genotype to Phenotype: The Big Picture

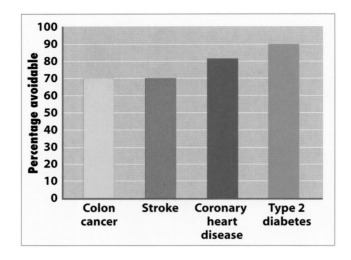

Biology Matters How to Avoid Life-Threatening Diseases

CHAPTER 11

Chromosomes and Human Genetics

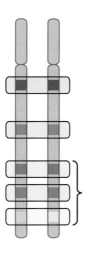

Figure 11.2 Genes Are Located on Chromosomes

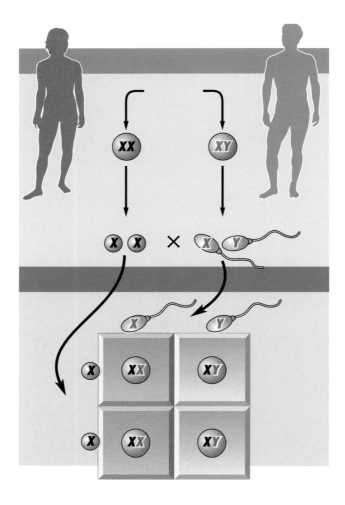

Figure 11.3 Sex Determination in Humans

If Your Blood Type Is:	You Can Give Blood To:	You Can Receive Blood From:
A+	A+, AB+	A+, A–, O+, O–
O+	O+, A+, B+, AB+	O+, O–
B+	B+, AB+	B+, B–, O+, O–
AB+	AB+	Everyone
A–	A+, A–, AB+, AB–	A–, O–
O–	Everyone	O–
B–	B+, B–, AB+, AB–	B–, O–
AB–	AB+, AB–	AB–, A–, B–, O–

Biology Matters Know Your Type

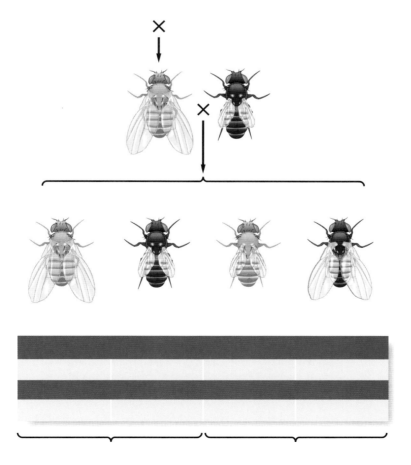

Figure 11.4 Some Alleles Do Not Assort Independently

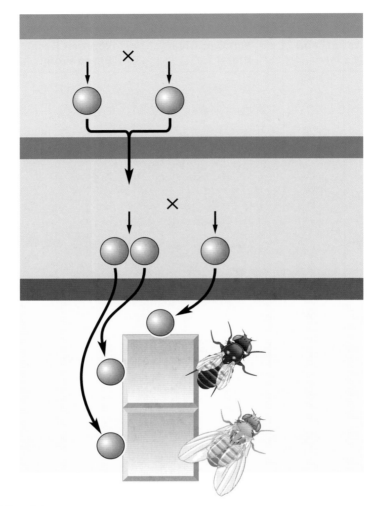

Figure 11.5 Linkage Is Not Complete

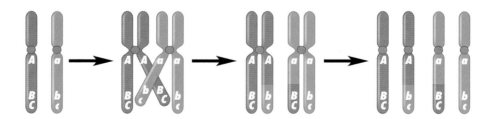

Figure 11.6 Crossing-Over Disrupts the Linkage Between Genes

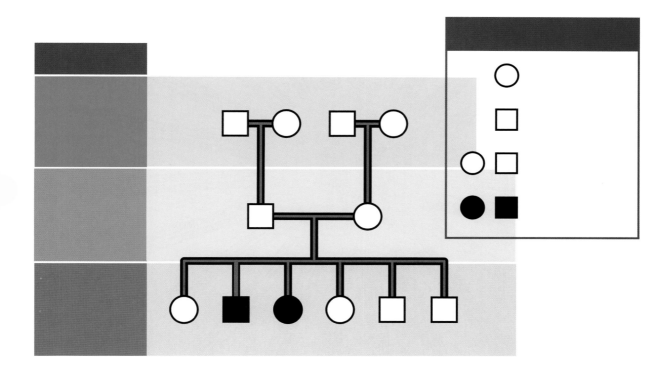

Figure 11.8 Human Pedigree Analysis

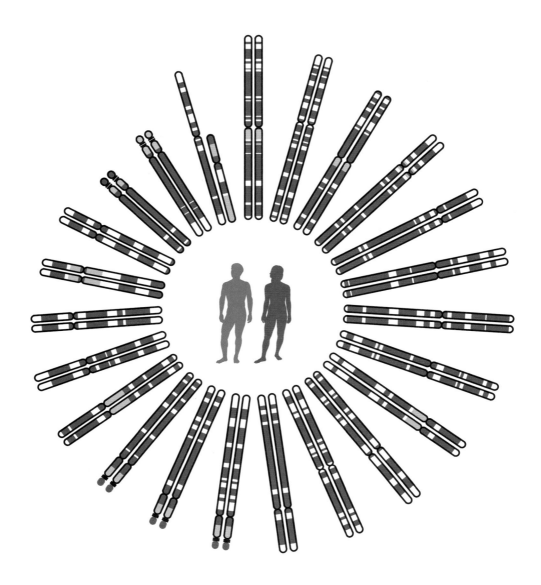

Figure 11.9 Examples of Single Genes That Cause Inherited Genetic Disorders

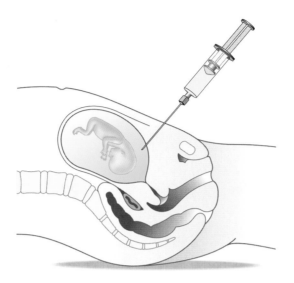

Science Toolkit Amniocentesis

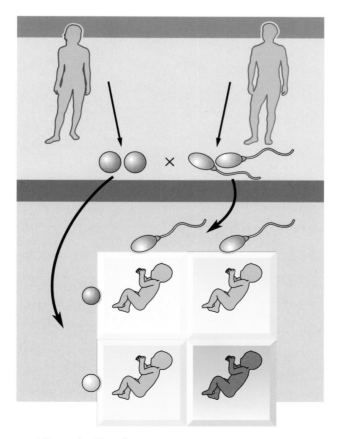

Figure 11.10 Inheritance of Autosomal Recessive Disorders

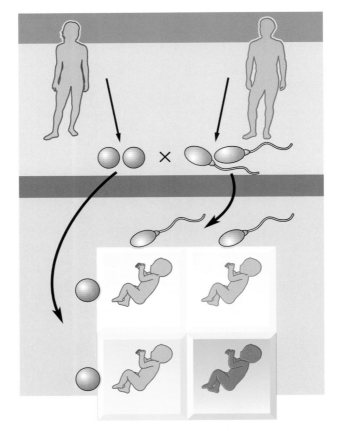

Figure 11.11 Inheritance of X-Linked Recessive Disorders

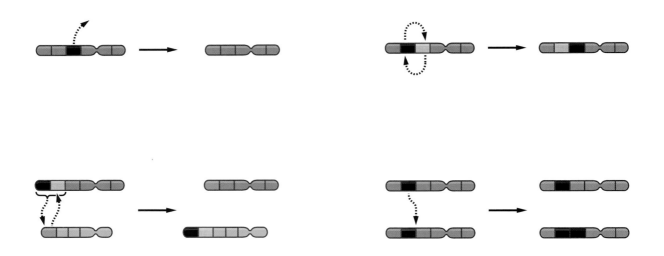

Figure 11.13 Structural Changes to Chromosomes

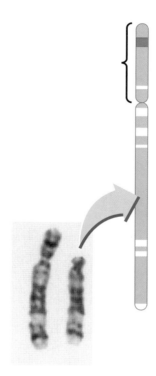

Figure 11.14 Cri du Chat Syndrome

Figure 12.1 Genetic Transformation of Bacteria

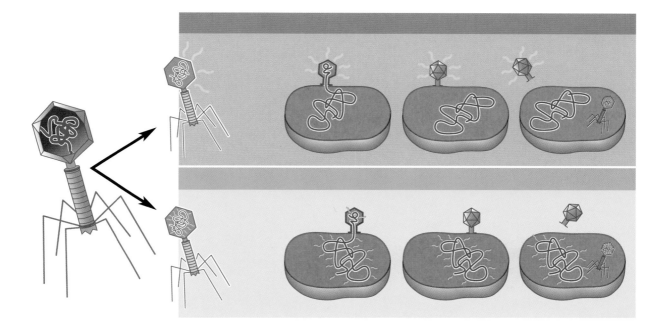

Figure 12.2 DNA Is the Genetic Material

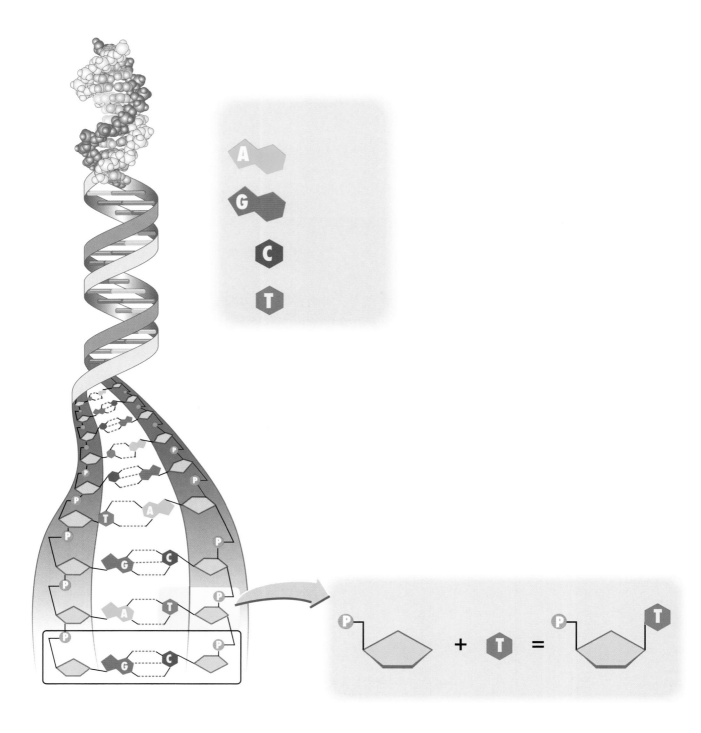

Figure 12.3 The DNA Double Helix and Its Building Blocks

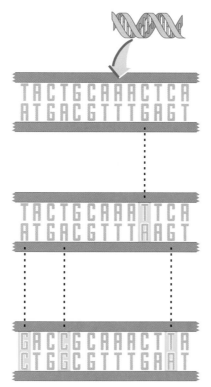

Figure 12.4 Variation in the Sequence of Bases in DNA

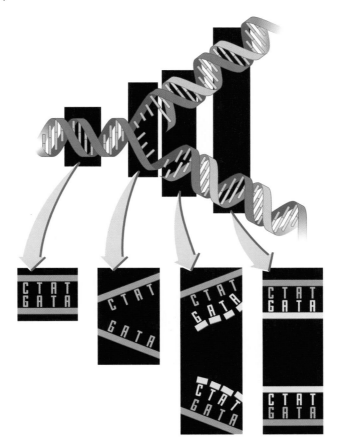

Figure 12.5 DNA Replication

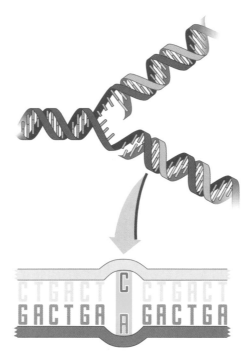

Figure 12.6 Mistakes Can Be Made in DNA Replication

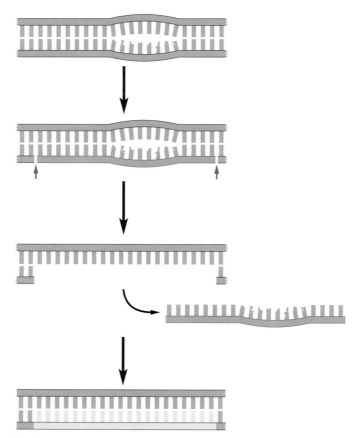

Figure 12.7 Repair Proteins Fix DNA Damage

Children's Average Exposure Relative to Adults

Air inhalation	3 to 1
Body surface area	2.25 to 1
Soil/dust consumption	3 to 1
Drinking water consumption	2.2 to 1
Dietary fat intake	3.4 to 1

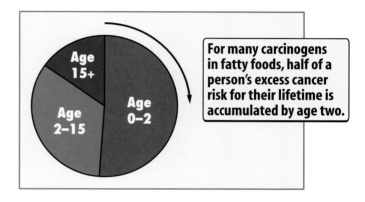

For many carcinogens in fatty foods, half of a person's excess cancer risk for their lifetime is accumulated by age two.

Some Mutagens That Damage DNA	Source of Exposure
Arsenic	• Arsenic pesticides in wooden playsets and decks • Drinking water (contaminants from mining and power plants)
Mutagen *X* and other by-products of water chlorination	• Drinking water
Formaldehyde	• Indoor air (offgases from building materials) • Paper, dyes, paper coatings
Benzene	• Gasoline fumes, glue, paint, furniture wax, detergent
PAHs (*polycyclic aromatic hydrocarbons*, a group of chemicals released by burning fossil fuels)	• Food and drinking water (contaminants from gasoline and coal-fired power plants)

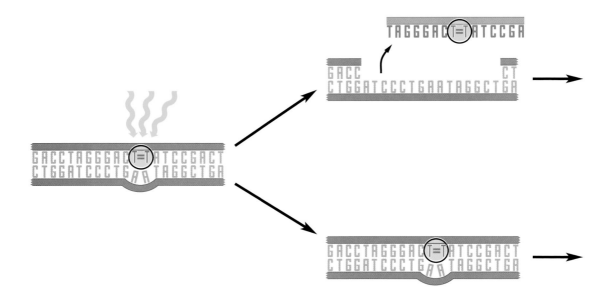

Figure 12.8 The Importance of DNA Repair Mechanisms

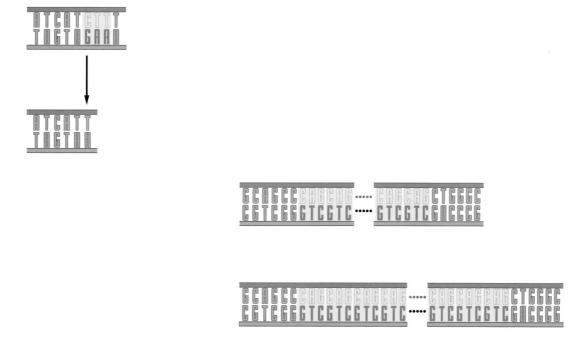

Figure 12.9 Two Deadly Disorders

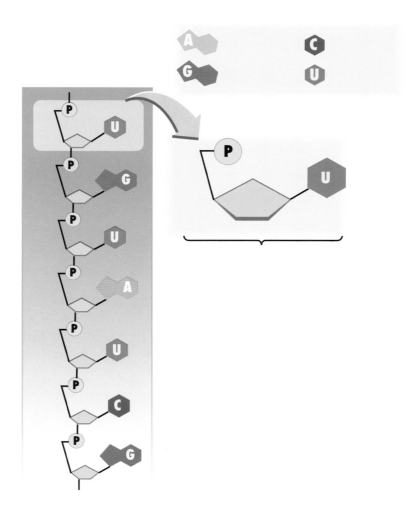

Figure 13.1 The Structure of RNA

Table 13.1

RNA Molecules and Their Functions

Type of RNA	Function	Shape
Messenger RNA (mRNA)	Specifies the order of amino acids in a protein	
Ribosomal RNA (rRNA)	Major component of ribosomes, the molecular machines that make the covalent bonds that link amino acids together into a protein	
Transfer RNA (tRNA)	Transports the correct amino acid to the ribosome, based on the information encoded in the mRNA	

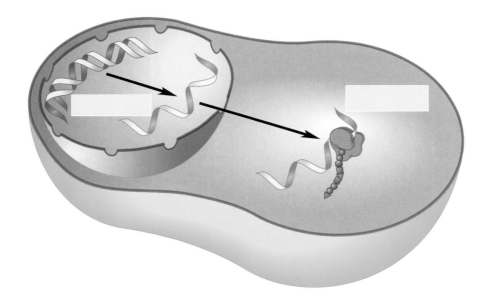

Figure 13.2 The Flow of Genetic Information in a Eukaryotic Cell

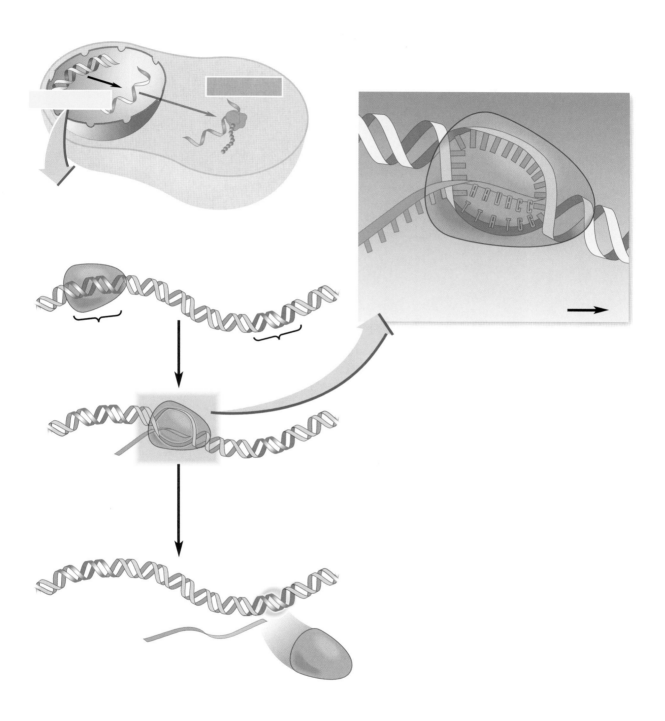

Figure 13.3 An Overview of Transcription

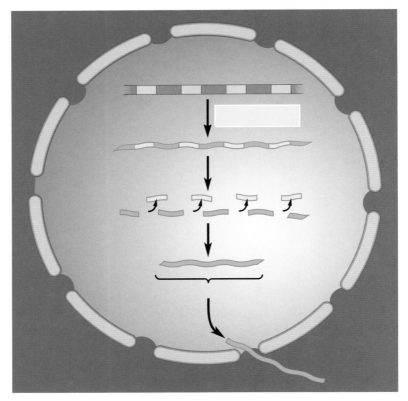

Figure 13.4 Removal of Introns by Eukaryotic Cells

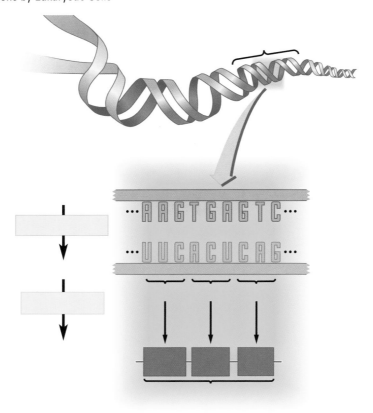

Figure 13.5 How Cells Use the Genetic Code

	U	C	A	G	
U	UUU UUC UUA UUG	UCU UCC UCA UCG	UAU UAC UAA UAG	UGU UGC UGA UGG	U C A G
C	CUU CUC CUA CUG	CCU CCC CCA CCG	CAU CAC CAA CAG	CGU CGC CGA CGG	U C A G
A	AUU AUC AUA AUG	ACU ACC ACA ACG	AAU AAC AAA AAG	AGU AGC AGA AGG	U C A G
G	GUU GUC GUA GUG	GCU GCC GCA GCG	GAU GAC GAA GAG	GGU GGC GGA GGG	U C A G

Figure 13.6 The Genetic Code

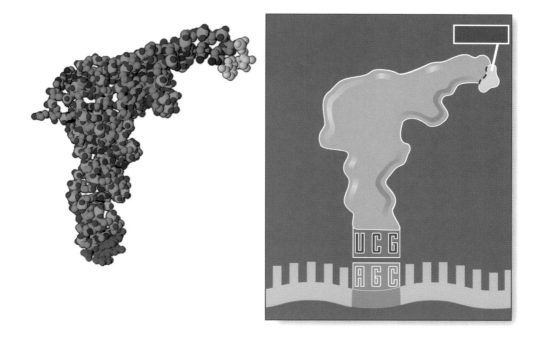

Figure 13.7 Transfer RNA (tRNA)

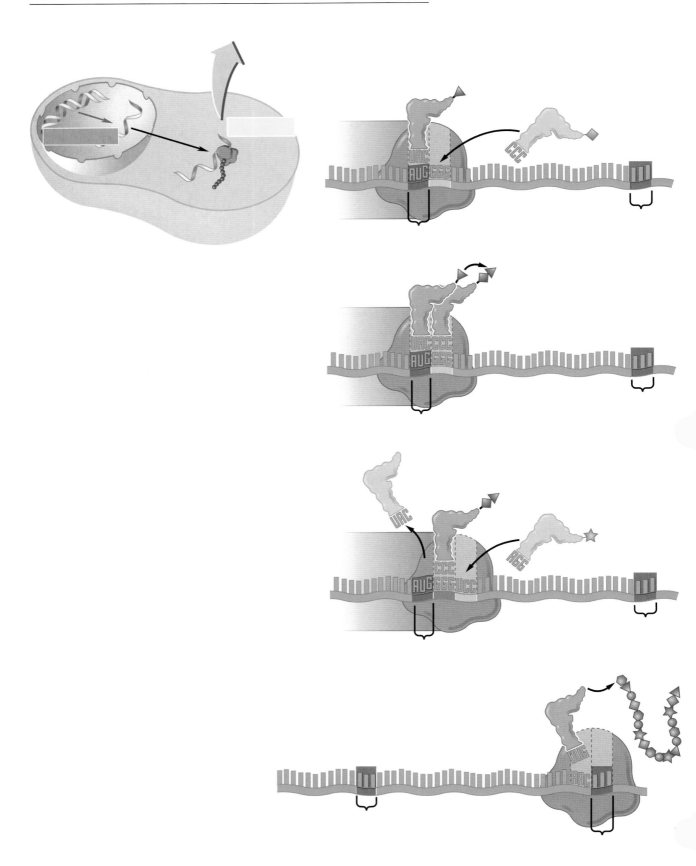

Figure 13.8 Translation

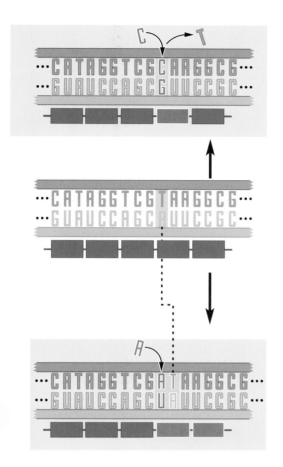

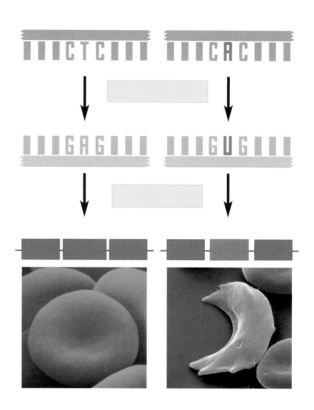

Figure 13.10 A Small Genetic Change Can Have a Large Effect

Figure 13.9 Effects of DNA Mutations on Protein Production

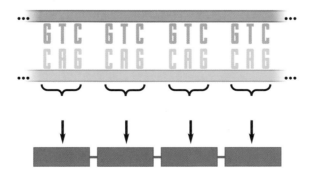

Figure 13.12 A Deadly Mutant

CHAPTER 14 Control of Gene Expression

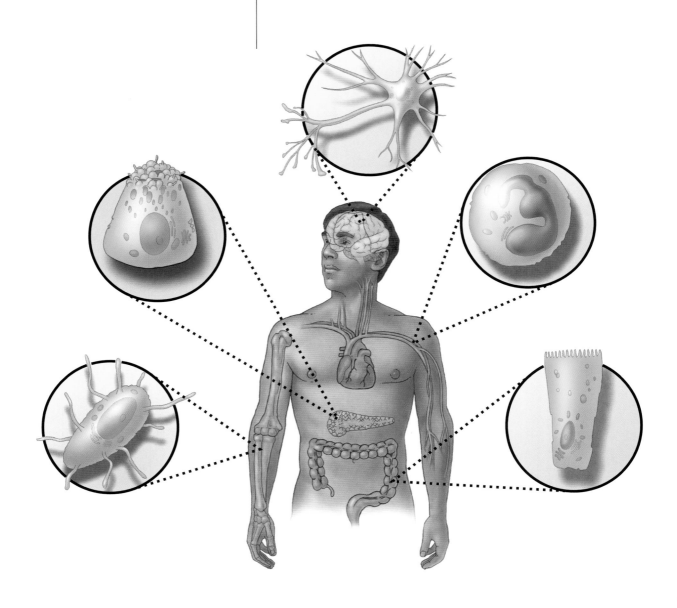

Figure 14.1 Different Cells Have the Same Genes

Figure 14.3 The Composition of Eukaryotic DNA

Table 14.1

Types of Eukaryotic DNA

Type	Subtypes	Description
Exons (of genes)		Most code for proteins used by the organism; others code for rRNAs, tRNAs, and various small RNAs that help regulate gene expression
Noncoding DNA	Introns	Sequences of noncoding DNA found within a gene that are removed from mRNA after transcription
	Spacer DNA	Sequences of noncoding DNA that separate genes
Transposons ("jumping genes")		Sequences of DNA that can move from one position on a chromosome to another, or from one chromosome to another

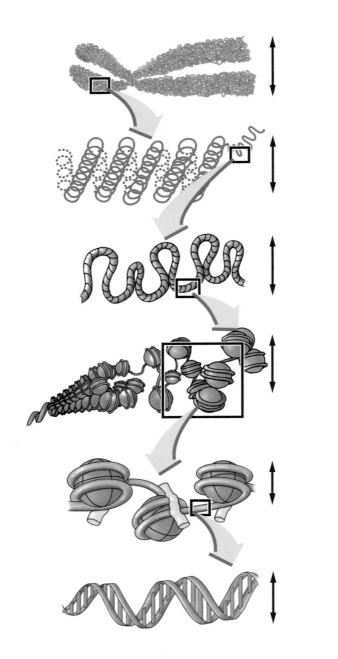

Figure 14.4 DNA Packing in Eukaryotes

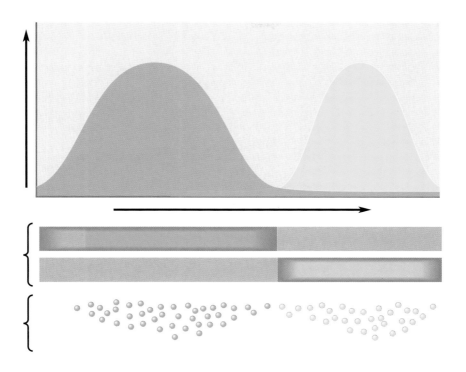

Figure 14.5 Bacteria Express Different Genes as Food Availability Changes

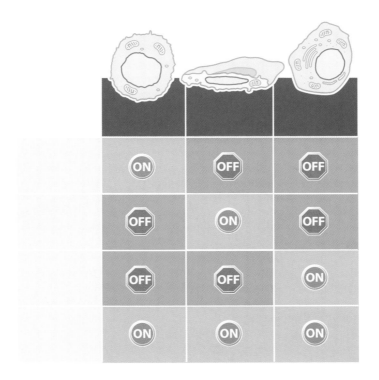

Figure 14.6 Different Types of Cells Express Different Genes

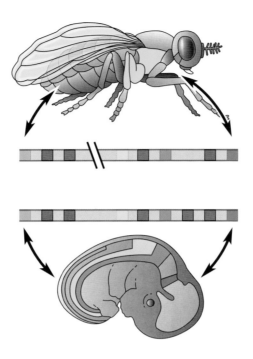

Figure 14.8 Homeotic Genes in Different Organisms Are Similar

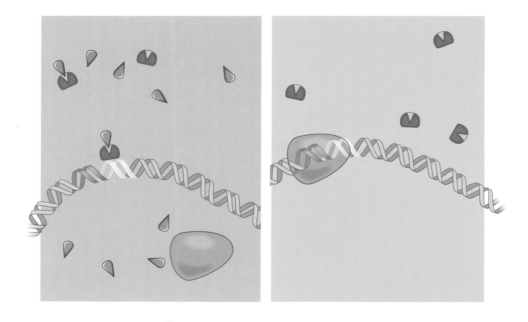

Figure 14.9 Repressor Proteins Turn Genes Off

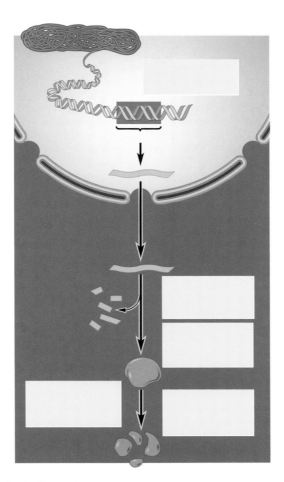

Figure 14.10 Control of Gene Expression in Eukaryotes

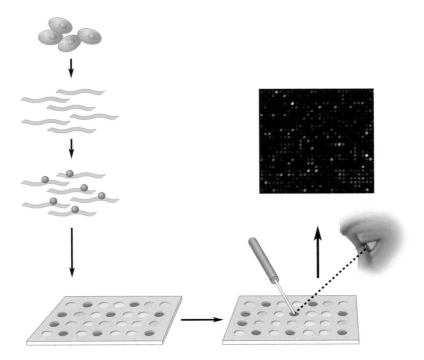

Science Toolkit Using DNA Chips

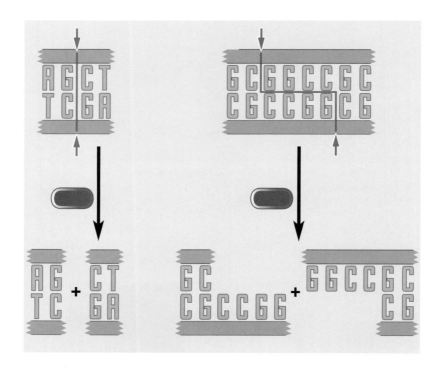

Figure 15.1 Restriction Enzymes Cut DNA at Specific Places

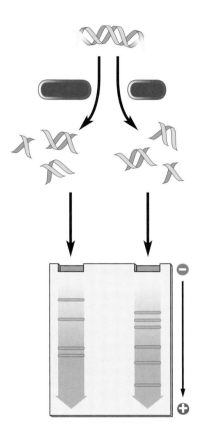

Figure 15.2 Gel Electrophoresis

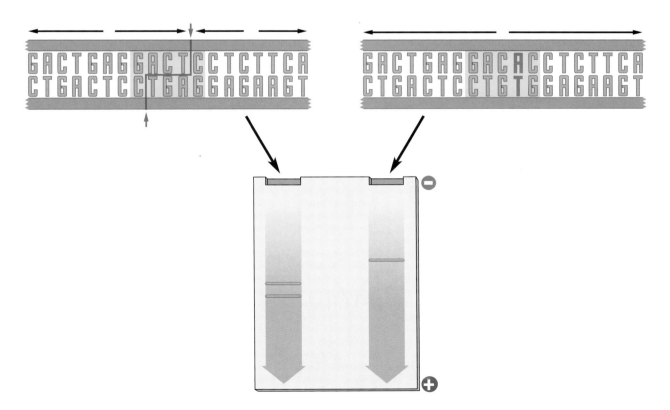

Figure 15.3 Identifying the Sickle-Cell Allele with Restriction Enzymes and Gel Electrophoresis

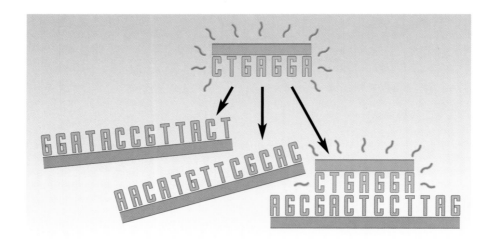

Figure 15.4 DNA Hybridization

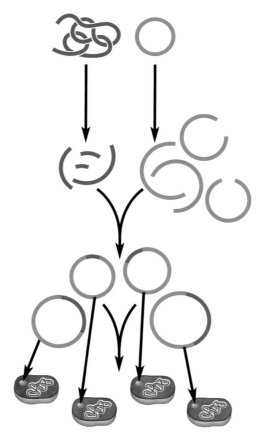

Figure 15.7 Construction of a DNA Library

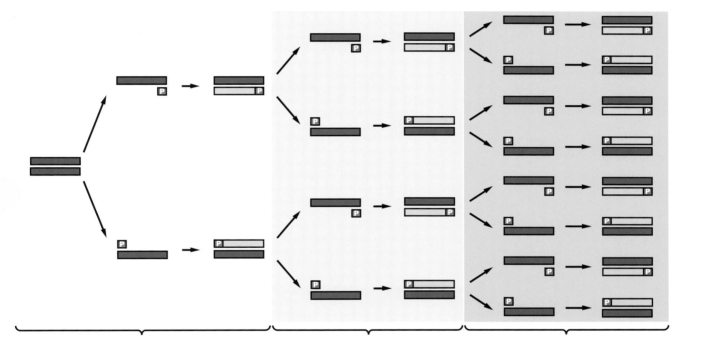

Figure 15.8 The Polymerase Chain Reaction

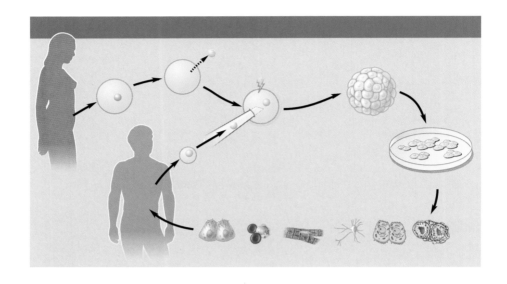

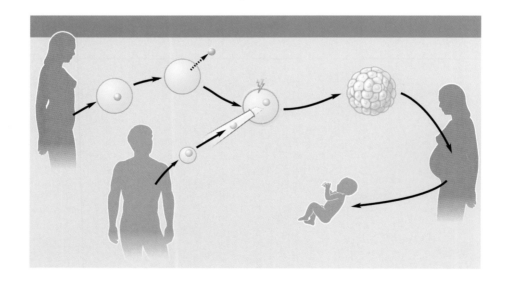

Science Toolkit Human Cloning

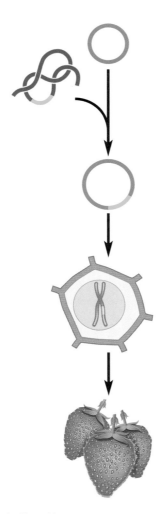

Figure 15.10 Genetic Engineering of Plants via Plasmids

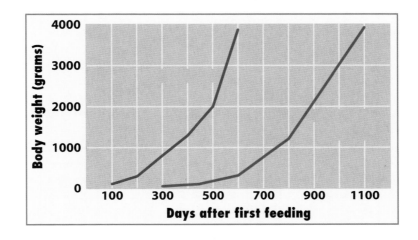

Figure 15.11 Feed Me!

Table 15.1

Methods of Production and Uses for Some Products of Genetic Engineering

For each product, the gene or DNA sequence that codes for the product is either inserted into host cells, such as *E. coli* or mammalian cells, or used in one of several automated procedures, such as a DNA synthesis machine or PCR. These cells or automated procedures are then used to make many copies of the product.

Product	Method of Production	Use
PROTEINS		
Human insulin	*E. coli*	Treatment of diabetes
Human growth hormone	*E. coli*	Treatment of growth disorders
Taxol biosynthesis enzyme	*E. coli*	Treatment of ovarian cancer
Luciferase (from firefly)	Bacterial cells	Testing for antibiotic resistance
Human clotting factor VIII	Mammalian cells	Treatment of hemophilia
ADA	Human cells	Treatment of ADA deficiency
DNA SEQUENCES		
Sickle-cell probe	DNA synthesis machine	Testing for sickle-cell anemia
BRCA1 probe	DNA synthesis machine	Testing for breast cancer mutations
HD probe	*E. coli*	Testing for Huntington disease
M13 probe	*E. coli*, PCR	DNA fingerprinting in plants
33.6 and other probes	*E. coli*, PCR	DNA fingerprinting in humans

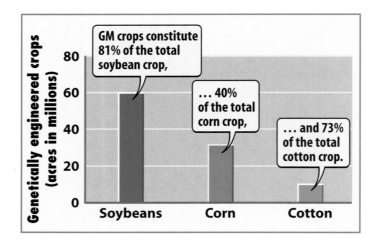

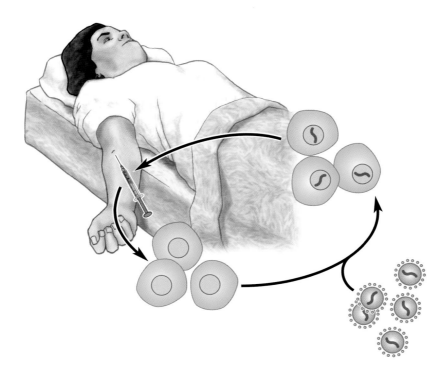

Figure 15.13 Gene Therapy for ADA Deficiency

Harnessing the Human Genome

Figure C.2 Milestones in the Quest for the Human Genome

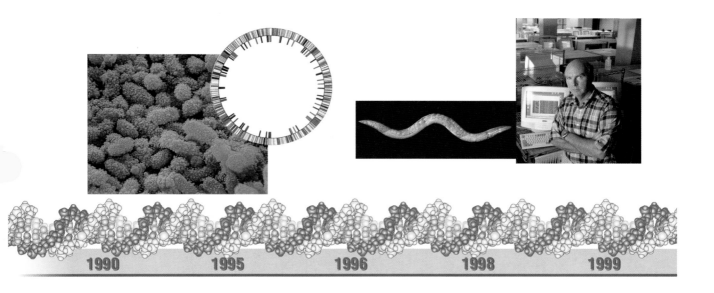

1990   1995   1996   1998   1999

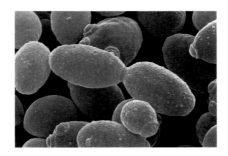

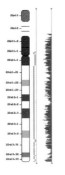

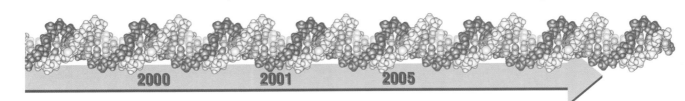

Figure C.2 Milestones in the Quest for the Human Genome (*continued*)

Table C.1

A Sample of the Genomes Sequenced to Date

Organism description	Scientific name	Date	Estimated genome size (millions of base pairs)	Predicted number of genes
Bacterium	*Haemophilus influenzae*	1995	1.8	1,740
Toxic shock bacterium	*Staphylococcus aureus*	2005	2.8	2,600
Bacterium	*E. coli*	1997	4.6	3,240
Anthrax bacterium	*Bacillus anthracis* [...an-THRASS-iss]	2003	5.2	5,000
Budding yeast	*Saccharomyces cerevisiae*	1996	12	6,000
Fruit fly	*Drosophila melanogaster*	2000	180	13,600
Nematode worm	*Caenorhabditis elegans*	1998	97	19,100

Table C.1

A Sample of the Genomes Sequenced to Date (*continued*)

Organism description	Scientific name	Date	Estimated genome size (millions of base pairs)	Predicted number of genes
Laboratory rat	*Rattus norvegicus* [...nor-VAY-juh-kuss]	2004	2,750	25,000
Flowering plant	*Arabidopsis thaliana* [uh-RAB-ih-DOP-siss THAH-lee-AH-nuh]	2000	125	25,500
Human	*Homo sapiens*	2001	3,200	20,000–30,000
Puffer fish	*Takifugu rubripes* [TAHK-ih-FOO-goo roo-BRIPE-eez]	2002	400	31,000

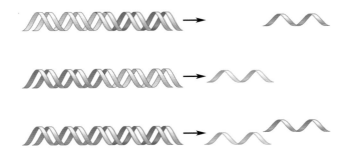

Figure C.3 How Genes Can Produce Multiple RNAs

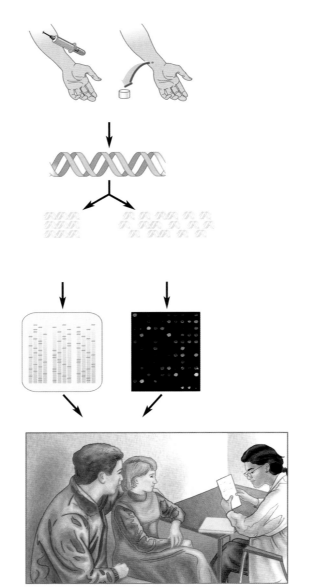

Figure C.6 From Tissue to Test Results

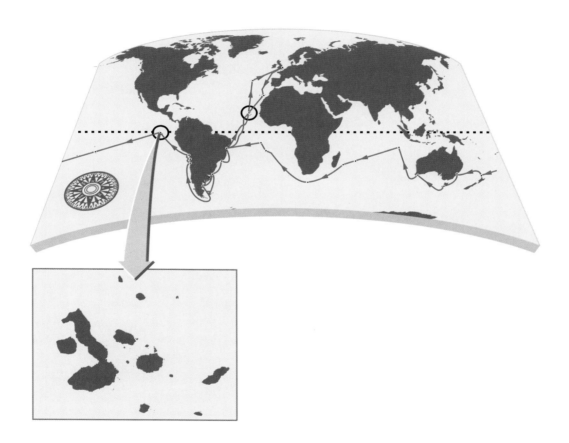

Chapter Opener Charles Darwin's Voyage

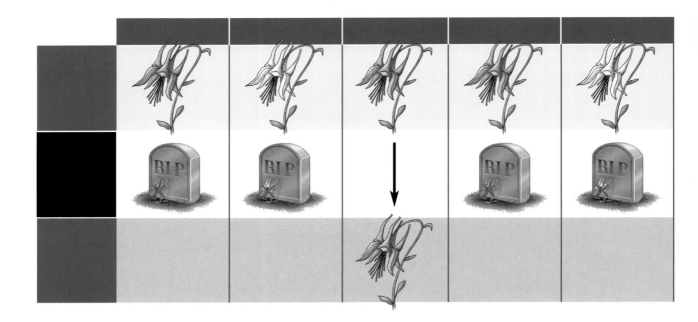

Figure 16.3 Genetic Drift

Figure 16.4 Shared Characteristics

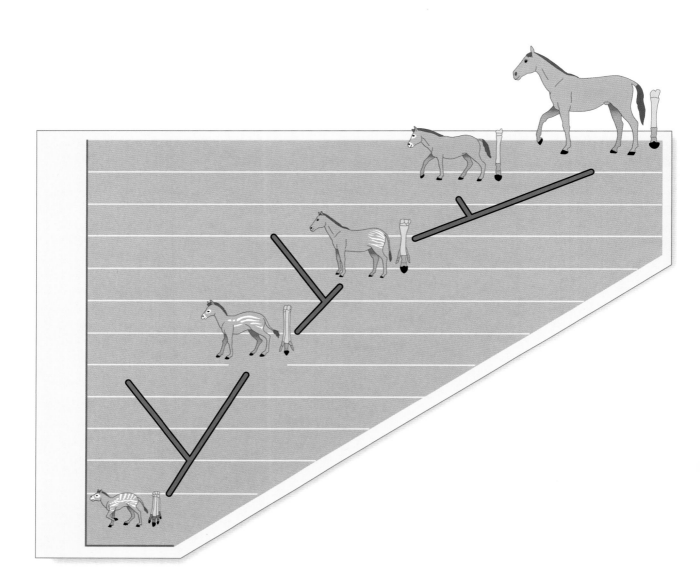

Figure 16.6 Fossil Evidence for Evolution

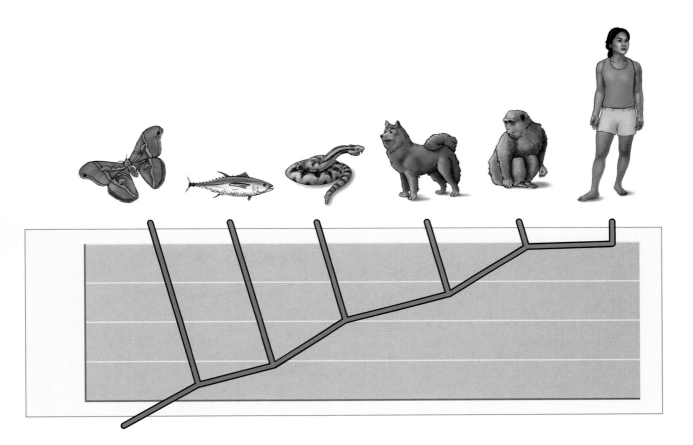

Figure 16.7 Independent Lines of Evidence Yield the Same Result

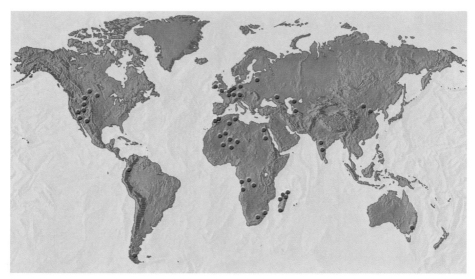

Figure 16.8 Once It Lived Throughout the Earth

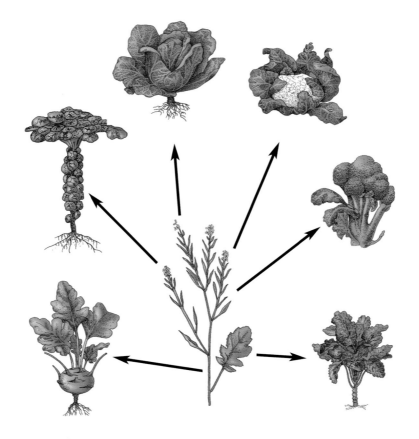

Figure 16.9 Artificial Selection Produces Genetic Change

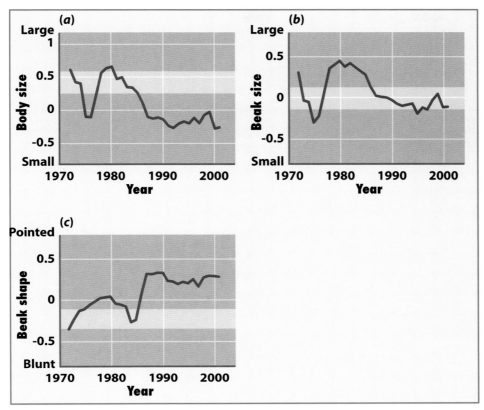

Figure 16.11 Evolutionary Change Within a Species

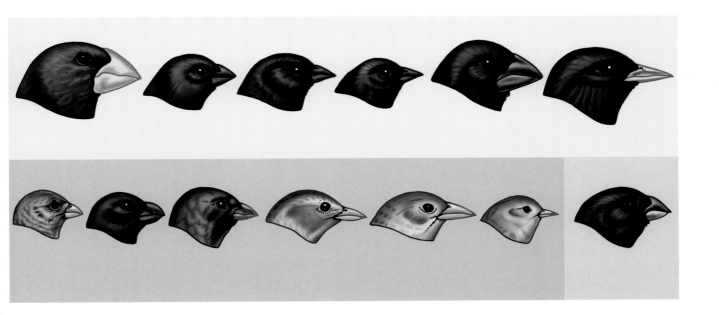

Figure 16.12 The Galápagos Finches

Figure 17.2 Gene Flow

$$p^2 + 2pq + q^2 = 1$$

Science Toolkit Testing Whether Evolution Is Occurring in Natural Populations

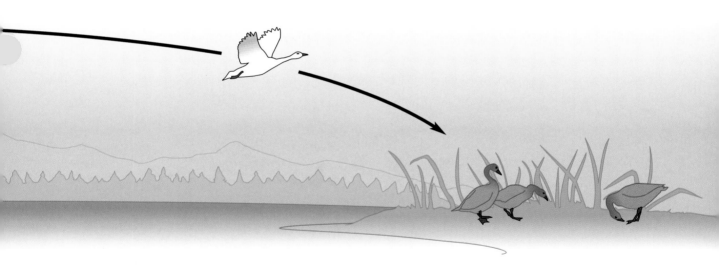

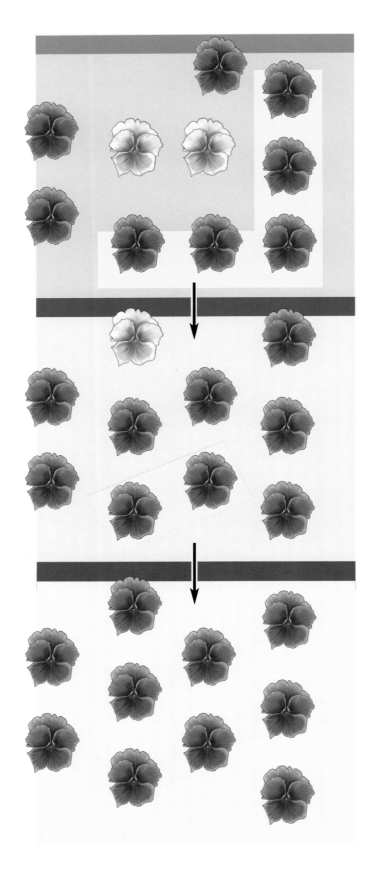

Figure 17.3 Genetic Drift

	Illinois		Kansas	Minnesota	Nebraska
	Prebottleneck (1933)	Postbottleneck	No bottleneck		
Population size					
No. of alleles at 6 genetic loci					
Percentage of eggs that hatch					

Figure 17.5 A Genetic Bottleneck

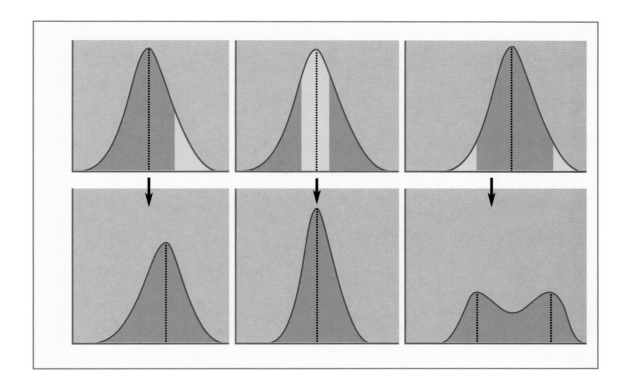

Figure 17.6 The Three Types of Natural Selection

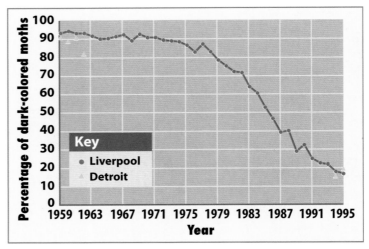

Figure 17.7 Directional Selection in the Peppered Moth

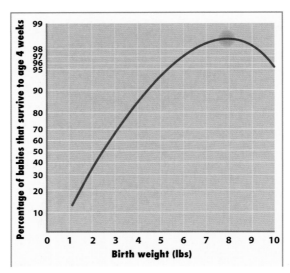

Figure 17.8 Stabilizing Selection for Human Birth Weight

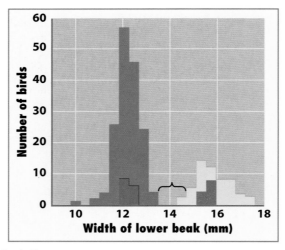

Figure 17.9 Disruptive Selection for Beak Size

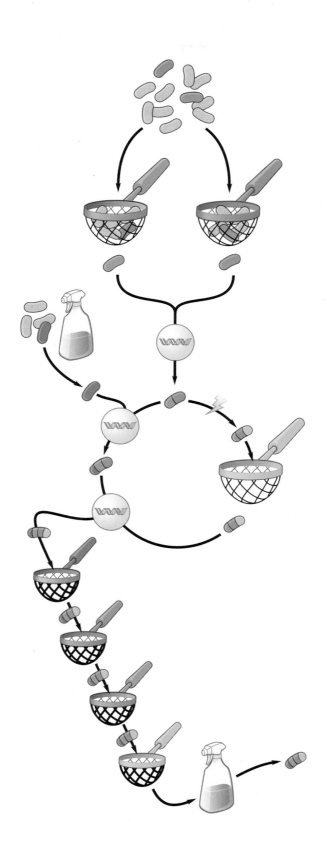

Figure 17.11 The Rise of Antibiotic Resistance

CHAPTER 18

Adaptation and Speciation

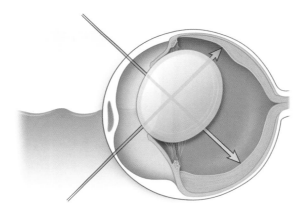

Figure 18.2 The Four-Eyed Fish, *Anableps anableps*

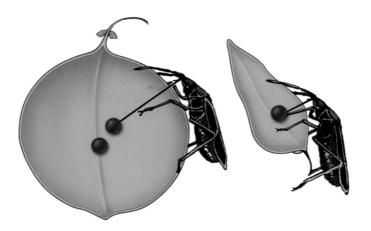

Figure 18.3 Rapid Evolution in an Insect

Table 18.1

Barriers That Can Reproductively Isolate Two Species in the Same Geographic Region		
Type of Barrier	**Description**	**Effect**
PREZYGOTIC BARRIERS		
Ecological isolation	The two species breed in different portions of their habitat, at different seasons, or at different times of the day	Mating is prevented
Behavioral isolation	The two species respond poorly to each other's courtship displays or other mating behaviors	Mating is prevented
Mechanical isolation	The two species are physically unable to mate	Mating is prevented
Gametic isolation	The gametes of the two species cannot fuse, or they survive poorly in the reproductive tract of the other species	Fertilization is prevented
POSTZYGOTIC BARRIERS		
Zygote death	Zygotes fail to develop properly and die before birth	No offspring are produced
Hybrid performance	Hybrids survive poorly or reproduce poorly	Hybrids are not successful

Figure 18.7 Some Species Interbreed Yet Remain Distinct

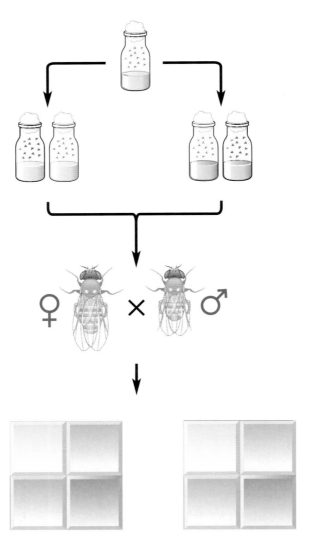

Figure 18.8 Selection Can Cause Reproductive Isolation

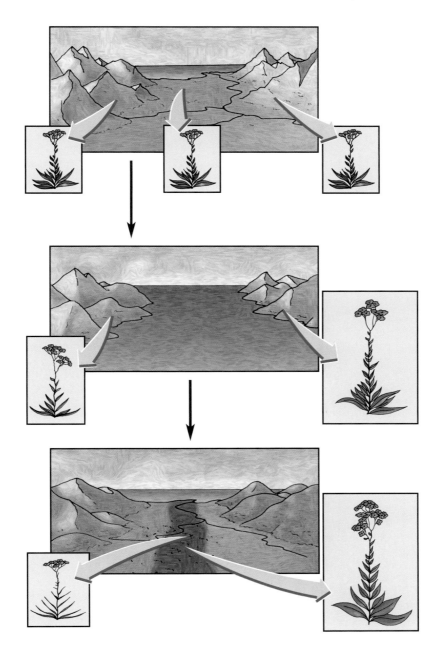

Figure 18.9 Allopatric Speciation

Figure 18.10 Lake Victoria Cichlids

19 The Evolutionary History of Life

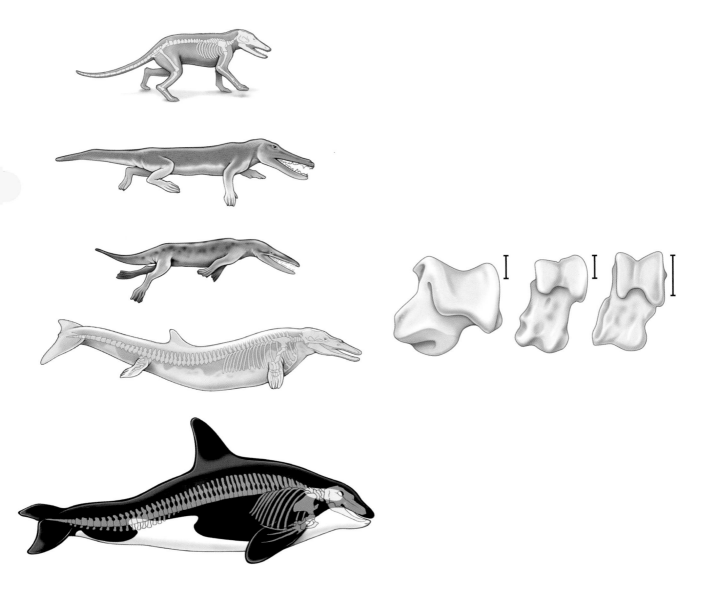

Figure 19.2 Shape-Shifters

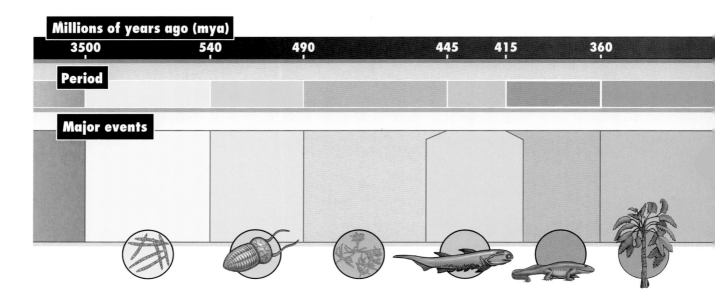

Figure 19.3 The History of Life on Earth: The Geologic Timescale

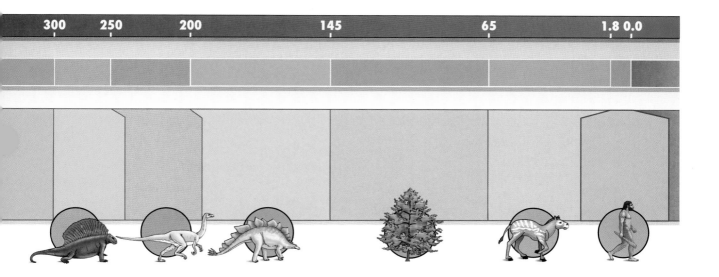

300   250   200   145   65   1.8 0.0

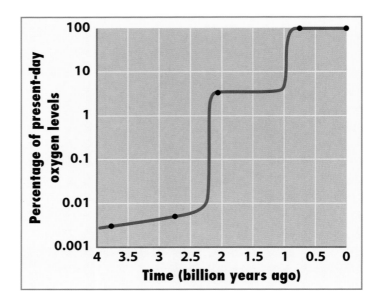

Figure 19.4 Oxygen on the Rise

Figure 19.5 Before and After the Cambrian Explosion

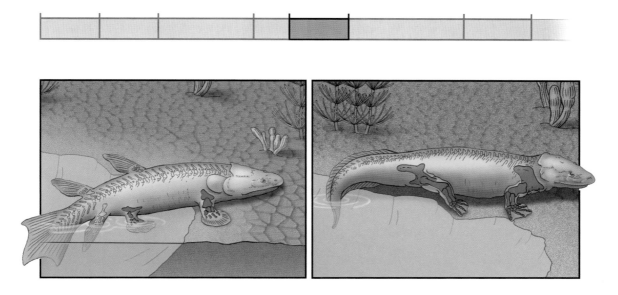

Figure 19.6 The First Amphibians

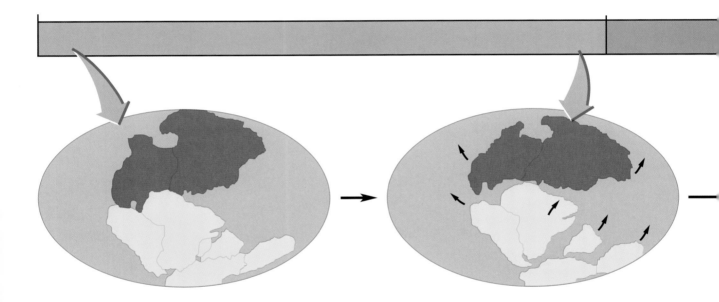

Figure 19.7 Movement of the Continents over Time

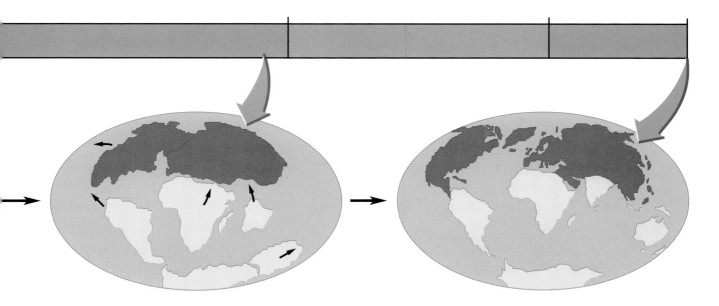

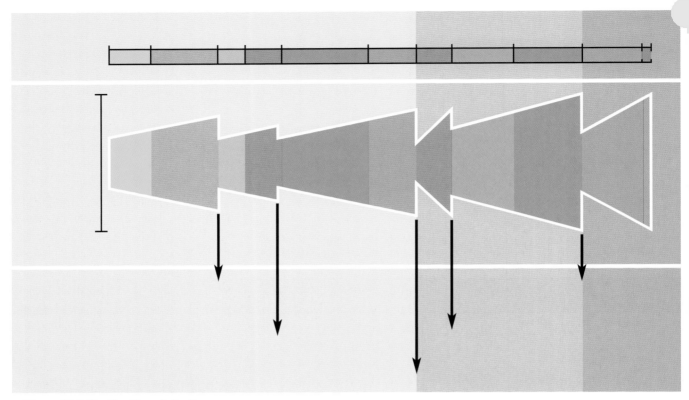

Figure 19.8 The Five Mass Extinctions

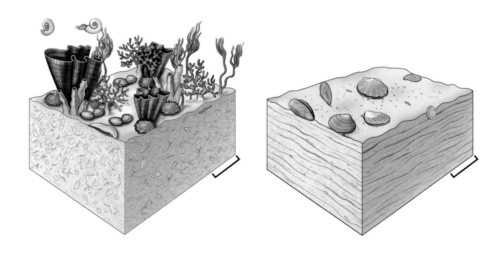

Figure 19.9 Before and After the Permian Mass Extinction

Biology Matters, Is a Mass Extinction Underway?

Table 1

Group	Number of Known Species	Number of Species Evaluated for Extinction Risk	Number of Threatened Species in 1996–1998	Number of Threatened Species in 2004	Threatened Species in 2004 (as a percentage of species evaluated)
VERTEBRATES					
Mammals	5,416	4,853	1,096	1,101	23 percent
Birds	9,917	9,917	1,107	1,213	12 percent
Amphibians	5,743	5,743	124	1,770	31 percent
PLANTS					
Gymnosperms	980	907	142	305	34 percent

Table 2

North America	Mammals	Birds	Reptiles	Amphibians	Fish	Mollusks	Other Invertebrates	Plants	Total
Canada	16	19	2	1	24	1	10	1	74
Saint Pierre and Miquelon	0	1	0	0	1	0	0	0	2
United States	40	71	27	50	154	261	300	240	1,143

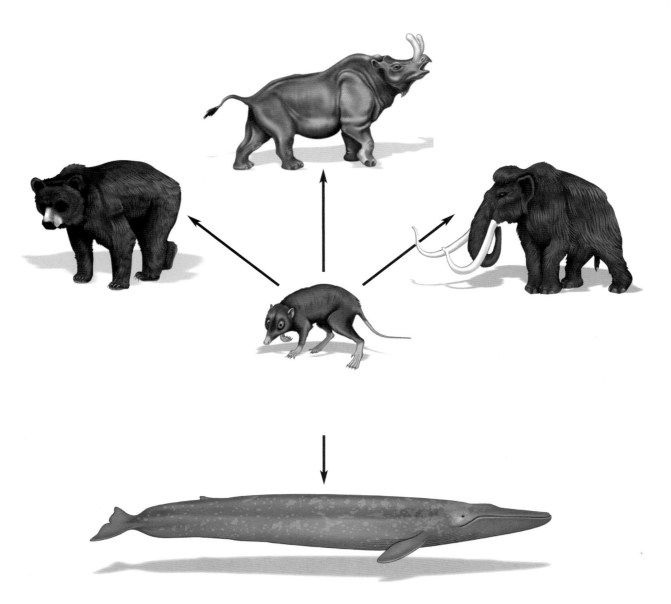

Figure 19.10 They Became Giants

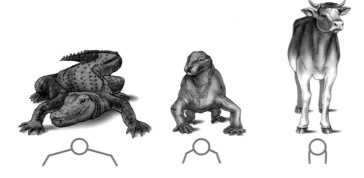

Figure 19.11 A Gradual Change in Gait

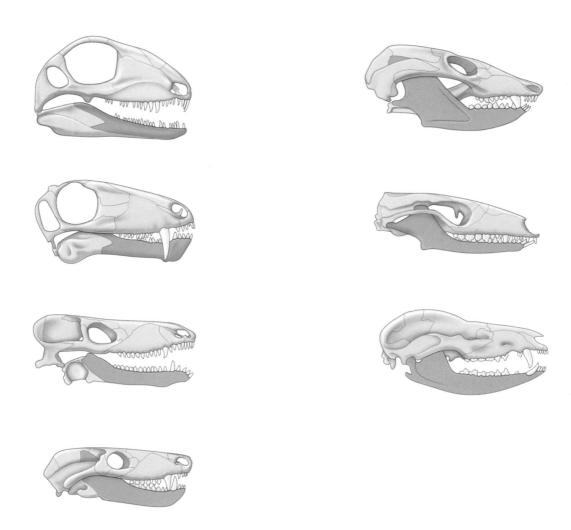

Figure 19.12 From Reptile to Mammal

Figure D.2 The Evolutionary Tree of Living Great Apes

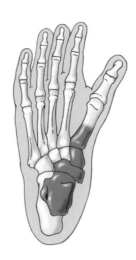

Figure D.4 Early Hominids Had Partially Opposable Big Toes

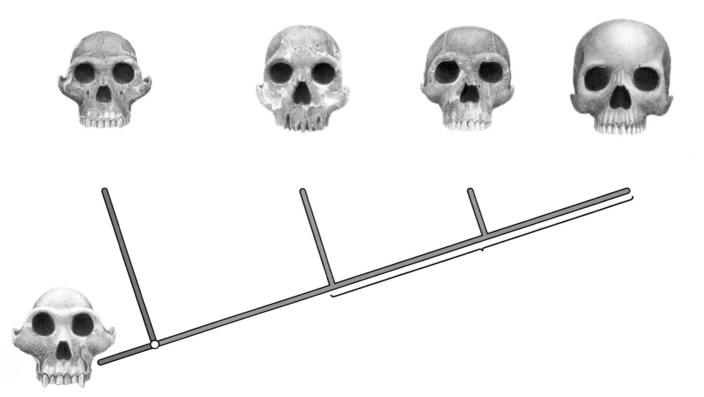

Figure D.6 A Gallery of Skulls

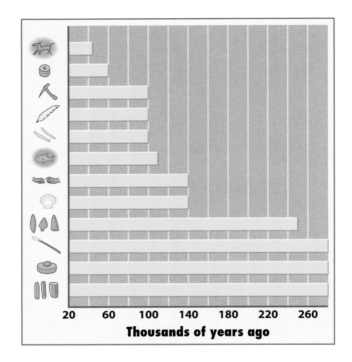

Figure D.7 Advanced Stone Age Tools and Art

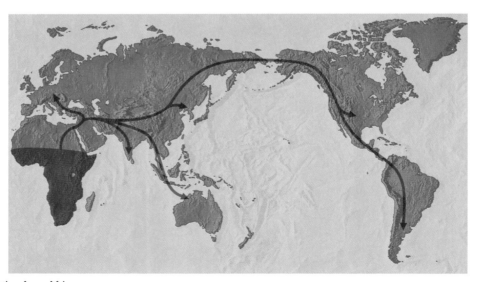

Figure D.8 Migration from Africa

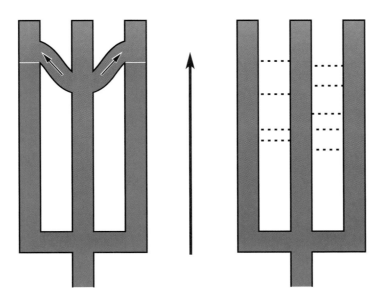

Figure D.9 The Origin of Anatomically Modern Humans

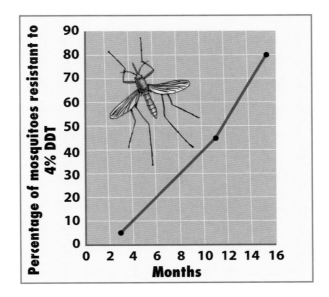

Figure D.11 Directional Selection for Pesticide Resistance

Some Areas of Increased Spending Due to Resistance	Annual Cost (in billions of dollars)
Additional pesticide use (to combat resistant insects)	1.2
Loss of crops	2–7
Treatment of patients infected with resistant *S. aureus*:	
Patients infected outside hospitals	14–21
Hospital infections: penicillin-resistant *S. aureus*	2–7
Hospital infections: methicillin-resistant *S. aureus*	8
HIV drug resistance	6.3
Total Cost	33–50

Biology Matters Humans: The World's Dominant Evolutionary Force?

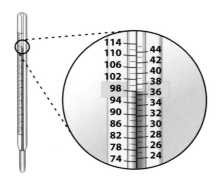

Figure 20.1 The Temperature of Life

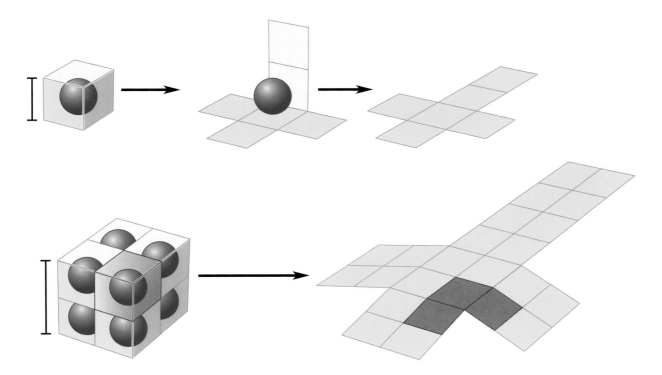

Figure 20.3 Size Affects Homeostasis

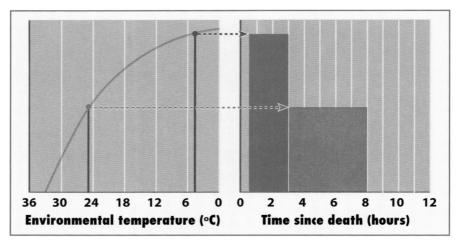

Figure 20.4 Crime Scene Investigators Can Use Loss of Body Heat to Estimate Time of Death

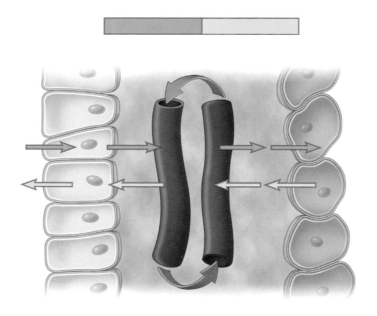

Figure 20.5 A Series of Regulated Exchanges

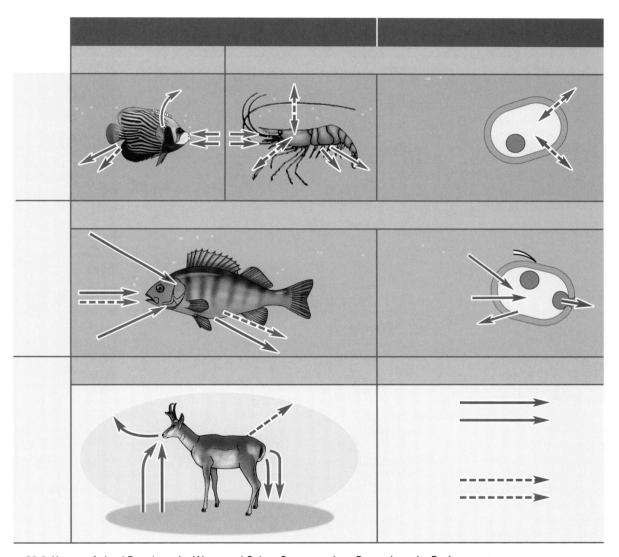

Figure 20.6 How an Animal Regulates Its Water and Solute Concentrations Depends on Its Environment

Table 20.1

A Comparison of the Solute Compositions of Marine Animal Body Fluid with That of Seawater and Fresh Water

Solute	Marine animal (lobster)	Seawater	Fresh water
Sodium	541	470	0.17
Potassium	8	10	Not detectable
Chloride	552	548	0.03

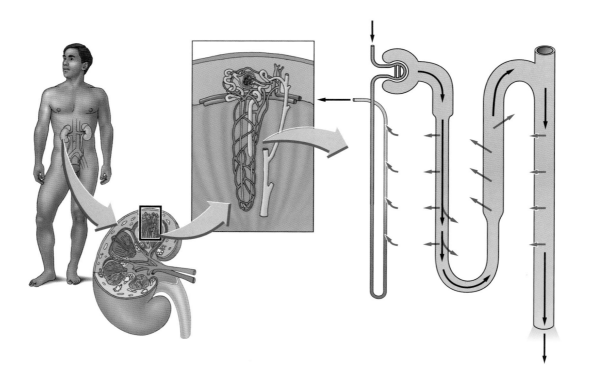

Figure 20.8 The Human Kidney Regulates Internal Water Content and Removes Toxic Wastes

Table 20.2

Substances Processed by the Human Kidney

	Substance	Percentage reabsorbed
NUTRIENTS	Glucose	100
	Sodium	99.4
	Chloride	99.1
WASTES	Urea (a waste product of the breakdown of proteins)	50
	Creatinine (a waste product produced by muscle tissue)	0

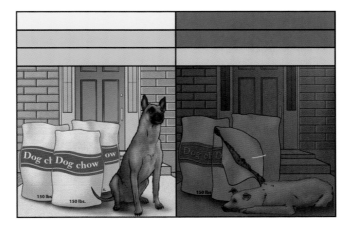

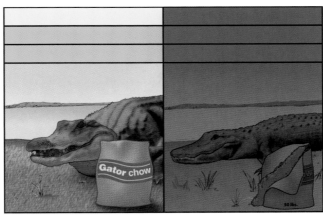

Figure 20.9 Temperature Regulation in Endotherms versus Ectotherms

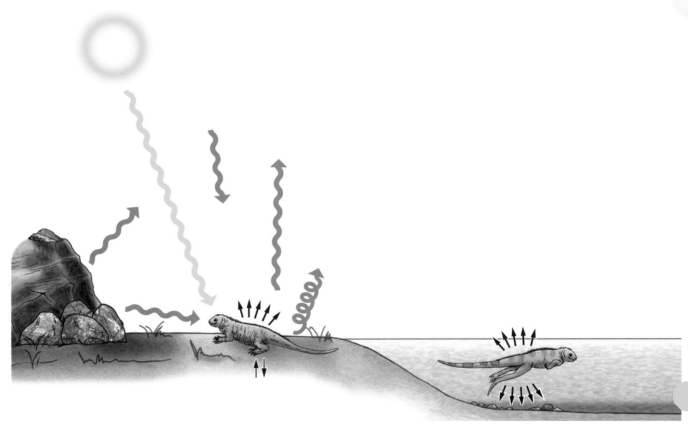

Figure 20.10 Heat Exchange on Land and in Water

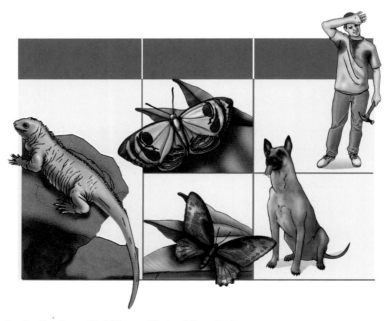

Figure 20.11 Some Strategies for Dealing with Different Kinds of Heat Exchanges

CHAPTER 21

Nutrition and Digestion

The Body Mass Index

		Normal						Overweight						Obese										Extreme obesity						
BMI	19	20	21	22	23	24	25	26	27	28	29	30	31	32	33	34	35	36	37	38	39	40	41	42	43	44	45	46
Height (inches)												Body weight (pounds)																
58	91	96	100	105	110	115	119	124	129	134	138	143	148	153	158	162	167	172	177	181	186	191	196	201	205	210	215	220
59	94	99	104	109	114	119	124	128	133	138	143	148	153	158	163	168	173	178	183	188	193	198	203	208	212	217	222	227
60	97	102	107	112	118	123	128	133	138	143	148	153	158	163	168	174	179	184	189	194	199	204	209	215	220	225	230	235
61	100	106	111	116	122	127	132	137	143	148	153	158	164	169	174	180	185	190	195	201	206	211	217	222	227	232	238	243
62	104	109	115	120	126	131	136	142	147	153	158	164	169	175	180	186	191	196	202	207	213	218	224	229	235	240	246	251
63	107	113	118	124	130	135	141	146	152	158	163	169	175	180	186	191	197	203	208	214	220	225	231	237	242	248	254	259
64	110	116	122	128	134	140	145	151	157	163	169	174	180	186	192	197	204	209	215	221	227	232	238	244	250	256	262	267
65	114	120	126	132	138	144	150	156	162	168	174	180	186	192	198	204	210	216	222	228	234	240	246	252	258	264	270	276
66	118	124	130	136	142	148	155	161	167	173	179	186	192	198	204	210	216	223	229	235	241	247	253	260	266	272	278	284
67	121	127	134	140	146	153	159	166	172	178	185	191	198	204	211	217	223	230	236	242	249	255	261	268	274	280	287	293
68	125	131	138	144	151	158	164	171	177	184	190	197	203	210	216	223	230	236	243	249	256	262	269	276	282	289	295	302
69	128	135	142	149	155	162	169	176	182	189	196	203	209	216	223	230	236	243	250	257	263	270	277	284	291	297	304	311

SOURCE: Adapted from *Clinical Guidelines on the Identification, Evaluation, and Treatment of Overweight and Obesity in Adults: The Evidence Report.*

	Percent of adults over 20 who are overweight	Percent of adults over 20 who are obese	Percent of adults over 20 who are severely obese
1999–2000	64.5	30.5	4.7
1988–1994	56.0	23.0	2.9
1976–1980	46.0	14.4	No data

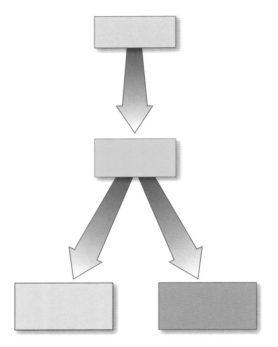

Figure 21.1 The Fate of Food: A Brief Overview

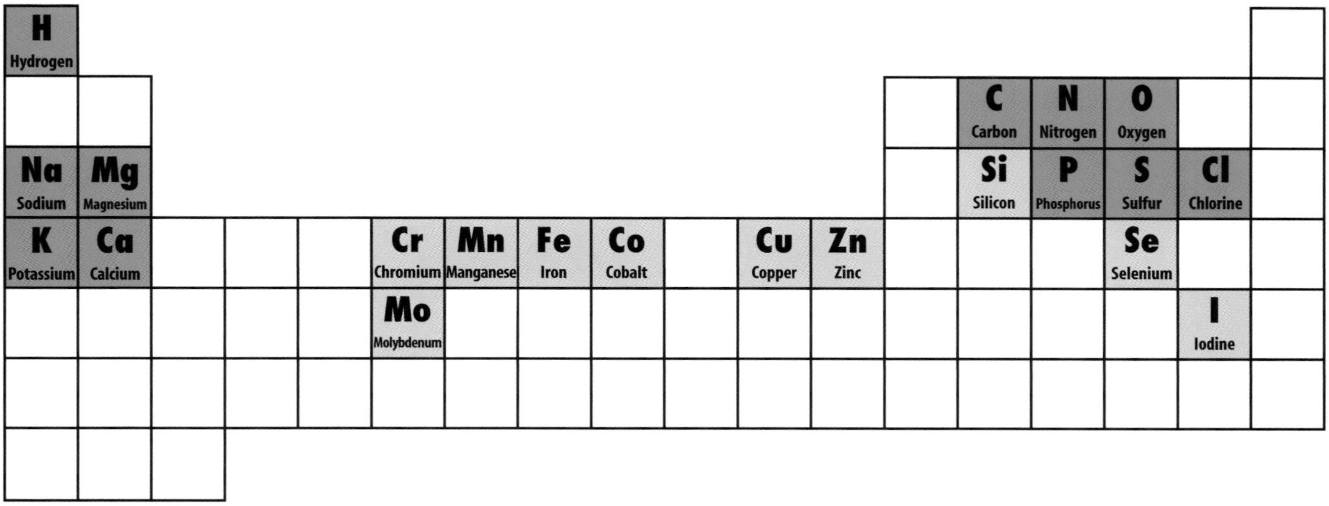

Figure 21.2 The Elements Essential to Life

Table 21.1

Nutrients Provide Animals with Energy

Note: **One calorie (abbreviated cal) = the energy needed to raise 1 gram of water 1°C. One kilocalorie (abbreviated kcal) = 1,000 calories, or the amount of energy needed to raise 1 kilogram of water 1°C. A 180-pound human can walk about a mile on 70 kcal.**

Nutrient	Absorbable units	Energy content of 1 gram (kcal)	Major use
Carbohydrates	Monosaccharides	4.8	Energy
Fats	Fatty acids, monoglycerides	9.6	Energy storage
Proteins	Amino acids	4.8	Building other proteins

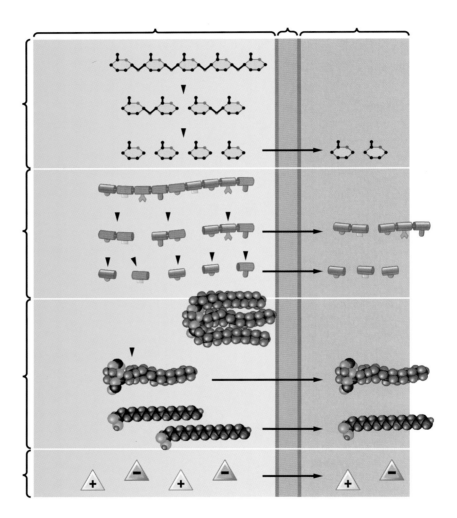

Figure 21.3 The Absorbable Forms of Nutrients

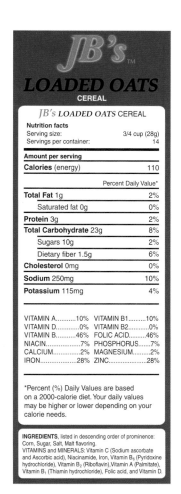

Figure 21.4 Nutritional Labels Can Tell Us a Great Deal

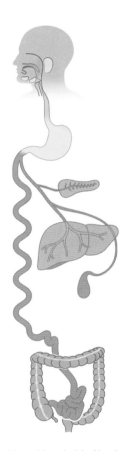

Figure 21.7 The Human Digestive System Converts Food into Absorbable Nutrients

Table 21.2

Vitamins Needed in the Human Diet

The human body cannot make these essential vitamins, or else makes them in insufficient amounts, and so must get what it needs from food. Vitamin C is a vitamin only for primates (including humans) and a few other animals (such as guinea pigs). All other animals can make vitamin C as needed.

Class	Vitamin	Main functions	Possible symptoms of deficiency and *excess*
Water-soluble	B vitamins: thiamine (B_1), riboflavin (B_2), niacin (nicotinamide), pyridoxine (B_6), pantothenic acid, folate (folic acid), cyanocobalamin (B_{12}), biotin	Act with enzymes to speed metabolic reactions, or act as raw materials for chemicals that do so. Work with enzymes to promote necessary biochemical reactions.	Deficiency: B vitamins act in concert; deficiency in one can cause symptoms related to deficiency in others. Deficiency diseases include pellagra and beriberi (damage to heart and muscles). *Excess: B_6 in excess can cause neurological damage.*
	Vitamin C (ascorbic acid)	Assists in maintenance of teeth, bones, and other tissues.	Deficiency: scurvy (teeth and bones degenerate), increased susceptibility to infection. *Excess: diarrhea, kidney stones with chronic overuse.*
Fat-soluble	Vitamin A (carotene)	Produces visual pigment needed for good eyesight; also used in making bone.	Deficiency: poor night vision; dry skin and hair. *Excess: nausea, vomiting, fragile bones.*
	Vitamin D	Promotes calcium absorption and bone formation.	Deficiency: poor formation of bones and teeth, irritability. *Excess: diarrhea and fatigue.*
	Vitamin E	Protects lipids in cell membranes and other cell components.	Deficiency: very rare. *Excess: heart problems.*
	Vitamin K	Produces clotting agent in the blood.	Deficiency: prolonged bleeding, slow wound healing. *Excess: liver damage.*

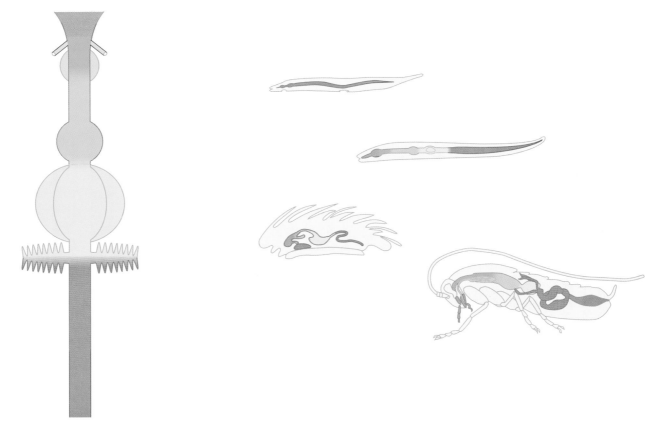

Figure 21.9 Animal Digestive Systems Share a Similar Organization

$$\text{Lactose} + H_2O \longrightarrow \text{Galactose} + \text{Glucose}$$

Figure 21.11 Lactase and Milk in the Human Diet

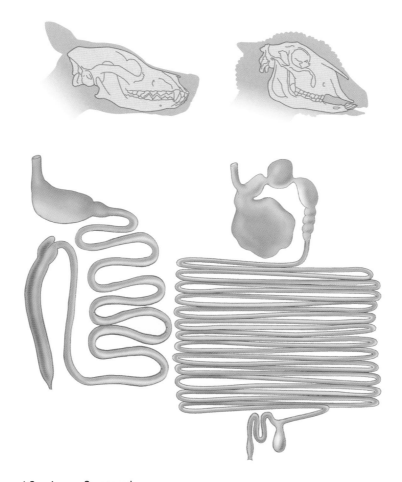

Figure 21.12 Herbivores and Carnivores Compared

CHAPTER 22   Gas Exchange

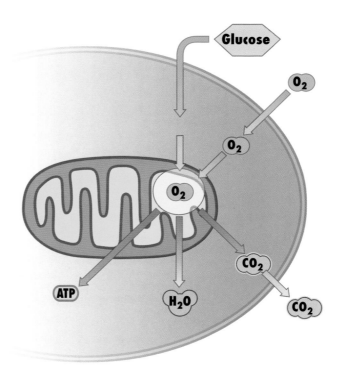

Figure 22.1 Respiration Occurs in Mitochondria

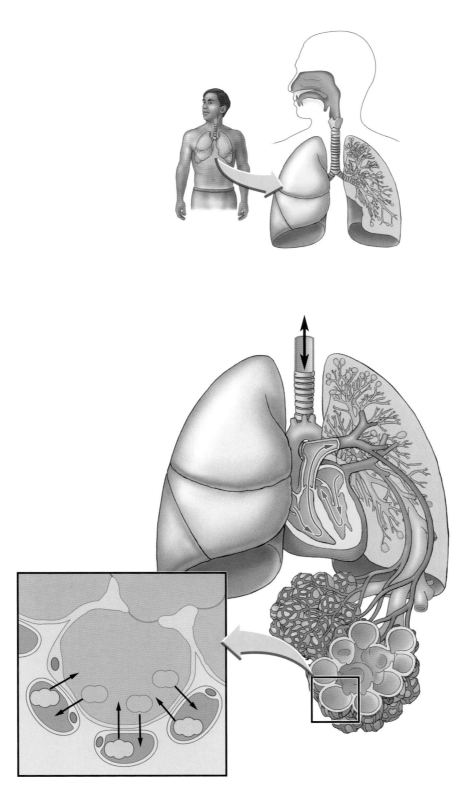

Figure 22.2 Gas Exchange in Humans

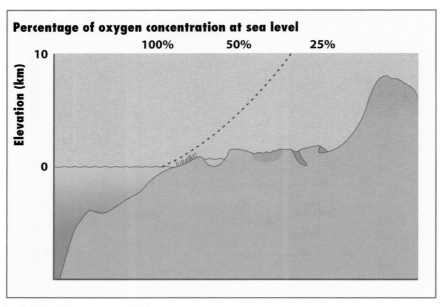

Figure 22.4 Oxygen Availability Differs among Habitats

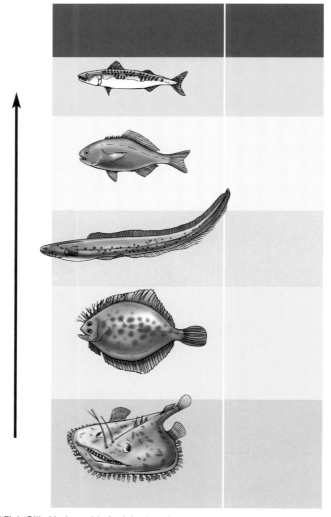

Figure 22.5 The Surface Area of Fish Gills Varies with Activity Level

Figure 22.6 The Relationship Between Body Size and Diffusion Time

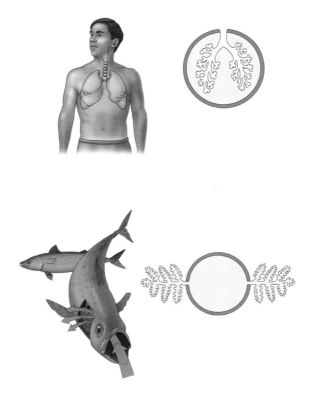

Figure 22.7 Lungs and Gills Compared

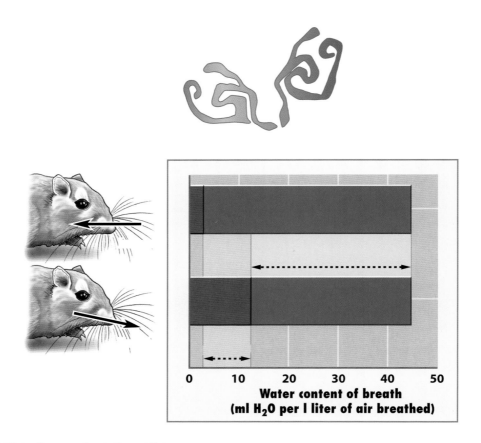

Figure 22.8 Water Conservation in Desert Rats

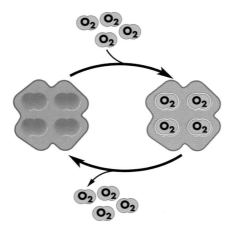

Figure 22.9 Hemoglobin Can Bind and Release Oxygen

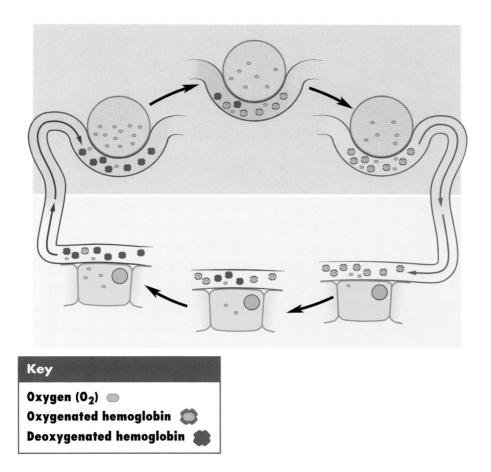

Key

Oxygen (O_2) ⬭
Oxygenated hemoglobin ✿
Deoxygenated hemoglobin ✿

Figure 22.10 How Hemoglobin Picks Up and Delivers Oxygen

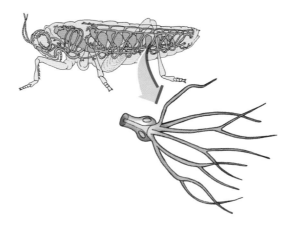

Figure 22.11 Insects "Breathe" Differently

CHAPTER 23 | Blood and Circulation

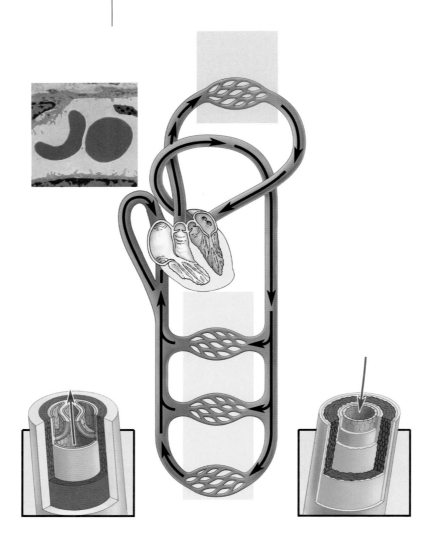

Figure 23.1 The Human Cardiovascular System

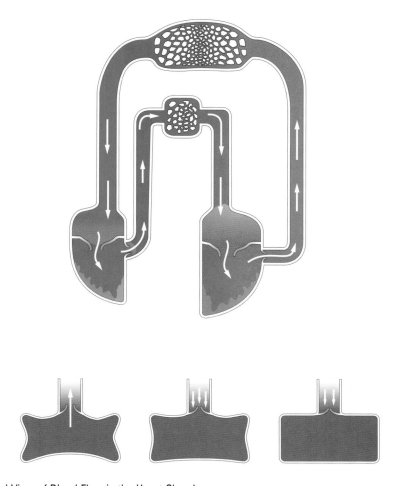

Figure 23.2 A Generalized View of Blood Flow in the Heart Chambers

Figure 23.3 The Human Heartbeat

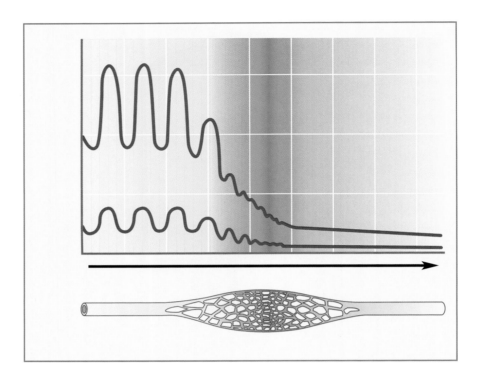

Figure 23.4 Pressure Changes in the Human Circulatory System

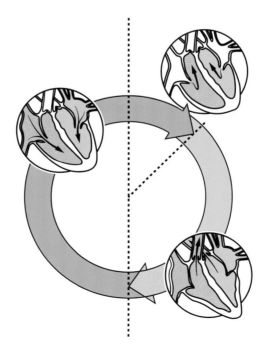

Figure 23.5 The Cardiac Cycle

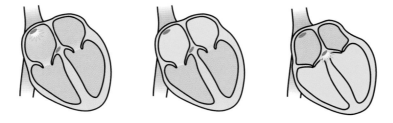

Figure 23.6 The Heart's Signaling System

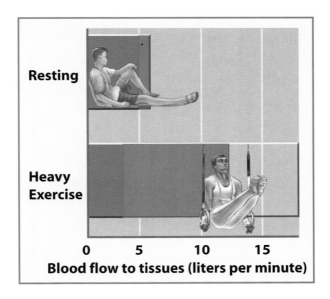

Figure 23.8 The Response of the Human Cardiovascular System to Exercise

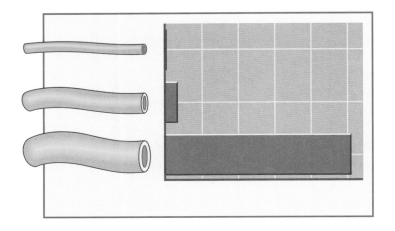

Figure 23.9 Blood Vessel Diameter and Frictional Drag

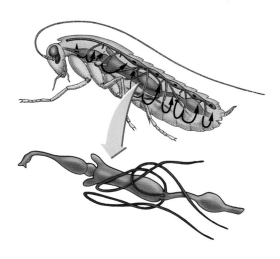

Figure 23.11 Open Circulatory Systems

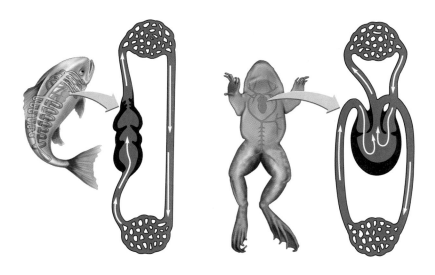

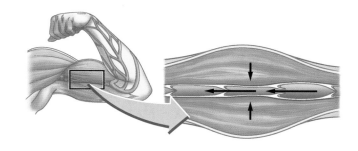

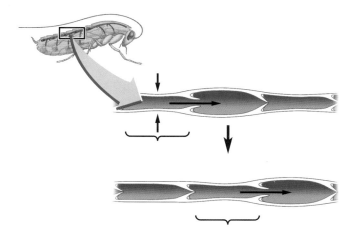

Figure 23.12 Some Other Types of Hearts and Pumps

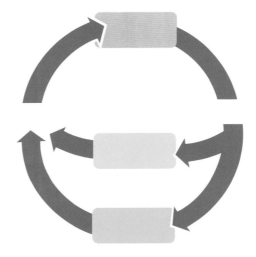

Figure 23.14 Arteries and Cholesterol: A Fatal Attraction

24 Animal Hormones

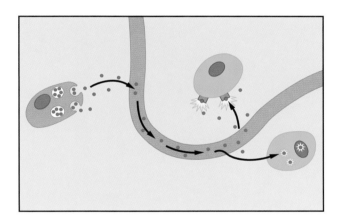

Figure 24.1 Hormones Allow Cells to Communicate with One Another

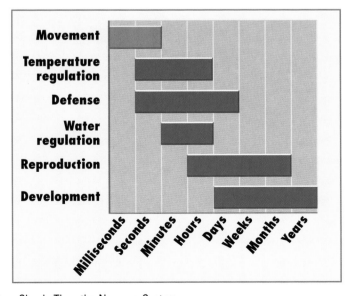

Figure 24.2 Hormones Act More Slowly Than the Nervous System

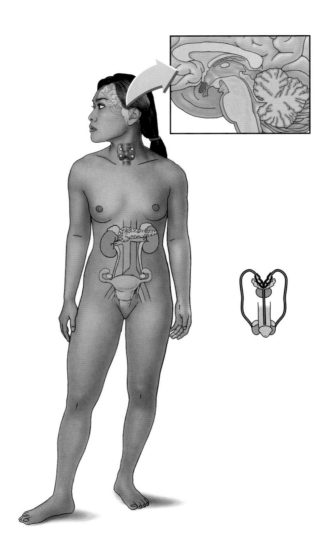

Figure 24.3 The Human Endocrine System

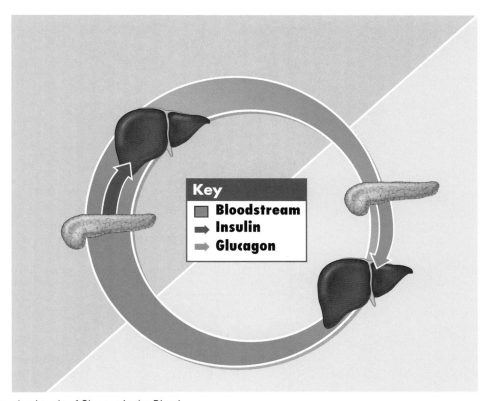

Figure 24.4 Balancing Levels of Glucose in the Blood

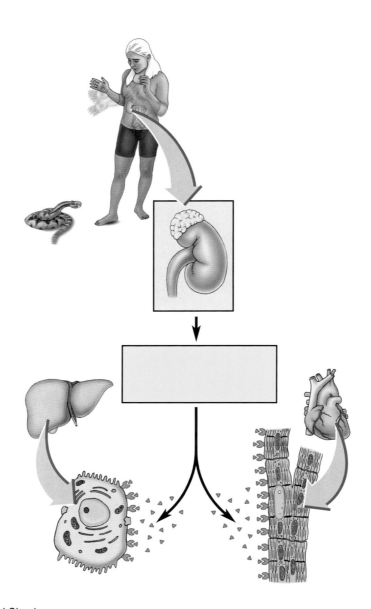

Figure 24.5 The Role of Adrenal Glands

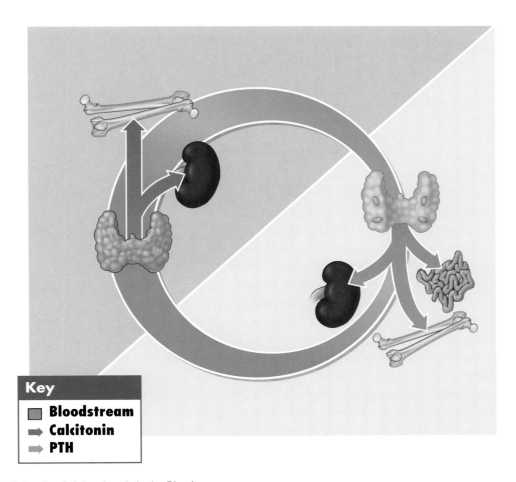

Key
- ⬜ **Bloodstream**
- ➡ **Calcitonin**
- ➡ **PTH**

Figure 24.7 Balancing Calcium Levels in the Blood

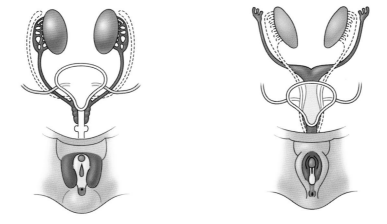

Figure 24.8 Sex-Specific Steroid Hormones

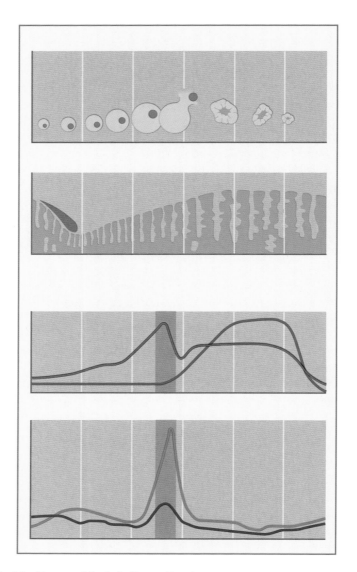

Figure 24.9 Hormonal Control of the Menstrual Cycle in Human Females

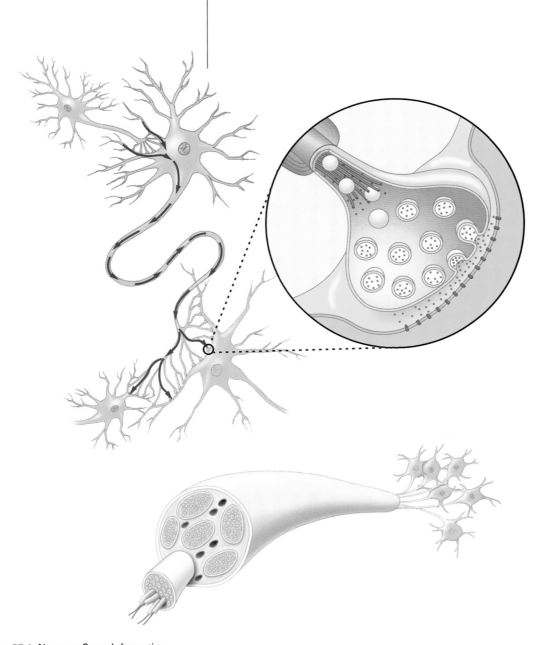

Figure 25.1 Neurons Carry Information

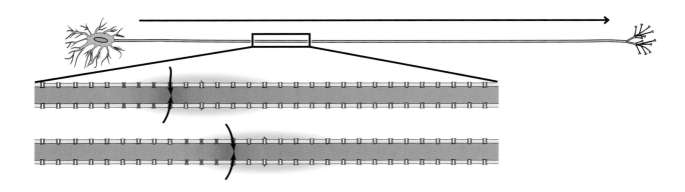

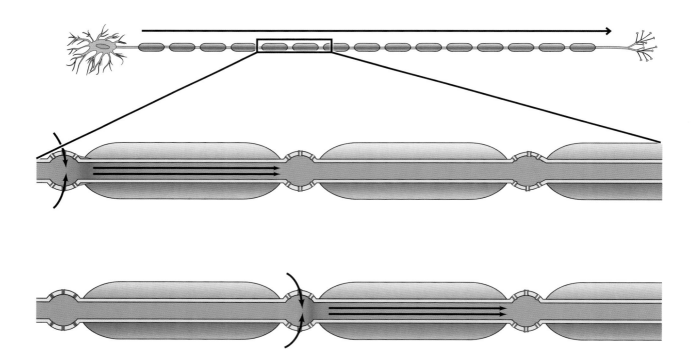

Figure 25.2 How Axons Transmit Signals

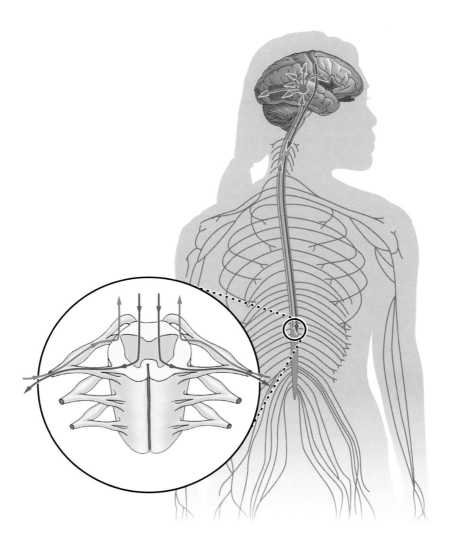

Figure 25.5 Organization of the Human Nervous System

Table 25.1

Functions of Common Human Neurotransmitters

Neurotransmitter	Major functions
Acetylcholine [ASS-uh-teel-KOH-leen]	Controls muscle contractions
Dopamine	Affects muscle activity; stimulates neurons in the pleasure center of the brain
GABA (gamma-aminobutyric acid) [. . . uh-MEEN-oh-byu-TEER-ik . . .]	Assists muscle coordination by inhibiting counterproductive or unneeded neurons
Serotonin [SAIR-uh-TOH-nin]	Regulates temperature, sleep, mood
Melatonin [MEL-uh-TOH-nin]	Helps regulate day-night cycles, including sleep
Enkephalins and endorphins	Inhibit transmission of signals from pain sensors, pain perception
Substance P	Regulates transmission of signals from pain sensors

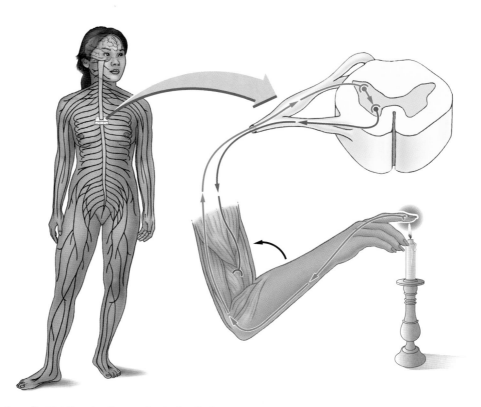

Figure 25.6 Reflex Arcs Do Not Require Processing in the Brain

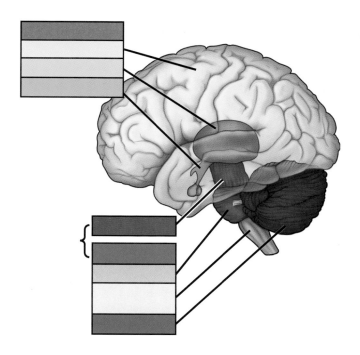

Figure 25.7 The Human Brain Has Three Specialized Regions

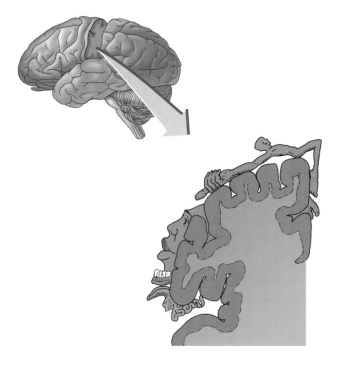

Figure 25.8 How the Brain "Sees" the Body

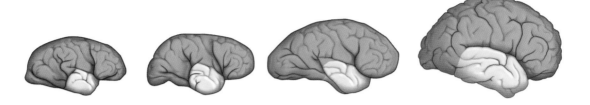

Figure 25.10 The Rise of the Neocortex

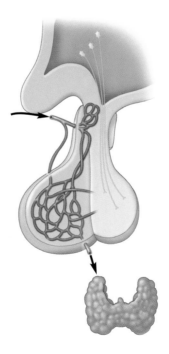

Figure 25.12 The Hypothalamus and Pituitary Gland Coordinate the Nervous and Endocrine Systems

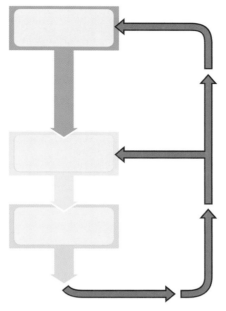

Figure 25.13 The Endocrine and Nervous Systems Work Together to Control the Rate at which Energy Is Released from Food

Table 25.2

Caffeine Content of Some Selected Items

Item	Typical content (mg)	Range (mg)
Coffee (8 oz. cup)		
Brewed, drip method	100	60–180
Instant	65	30–120
Decaffeinated	3	1–5
Tea (8 oz. cup)		
Brewed, major U.S. brands	40	20–90
Brewed, imported brands	60	25–110
Instant	28	24–31
Iced	25	9–50
Some soft drinks (8 oz.)	24	20–40
Cocoa (8 oz.)	6	3–32
Chocolate milk (8 oz.)	5	2–7
Milk chocolate (1 oz.)	6	1–15
Dark chocolate, semi-sweet (1 oz.)	20	5–35
Baker's chocolate (1 oz.)	26	26
Chocolate-flavored syrup (1 oz.)	4	4

26 Sensing the Environment

Table 26.1

Different Ways to Sense the World

The sensory receptors listed here are found in many animals, including most vertebrates. Those marked with an asterisk are not found in humans.

Receptor type	Stimulus	Sense
Photoreceptors	Light	Vision
Chemoreceptors	Chemicals	Taste
		Smell
Mechanoreceptors	Physical changes	Touch
		Hearing
		Proprioception (body position)
		Balance
		Thermoreception (gradations of heat and cold)
		Pain
*Electroreceptors	Electrical fields (especially those generated by muscle contractions of other animals)	*Electrical sense
*Magnetoreceptors	Magnetic fields (especially of Earth)	*Magnetic sense

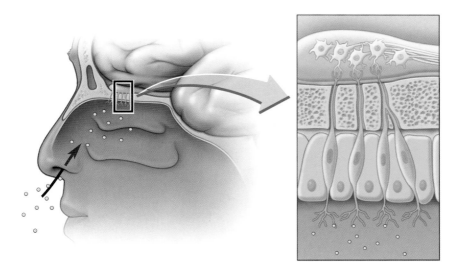

Figure 26.2 Smell in Humans

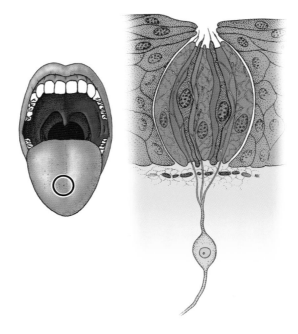

Figure 26.3 Taste in Humans

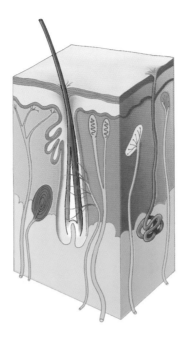

Figure 26.4 Some Mechanoreceptors in Human Skin

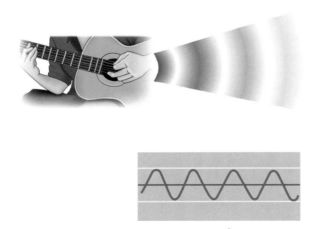

Figure 26.5 Sound Is Composed of Pressure Waves

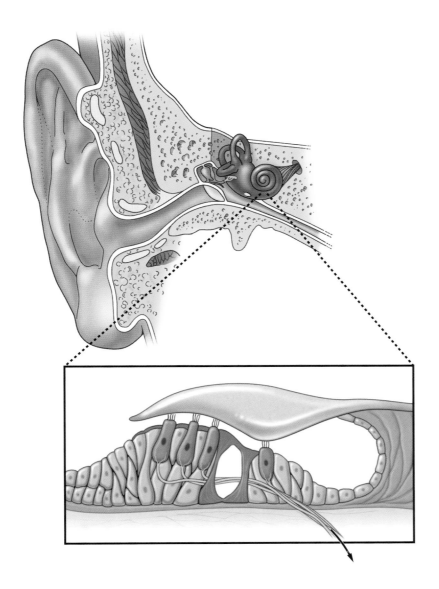

Figure 26.6 The Human Ear Detects Minute Pressure Changes

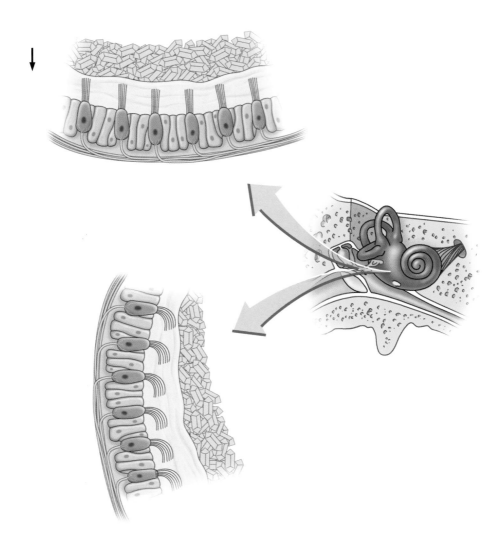

Figure 26.7 Our Ears Help Us Keep Our Balance

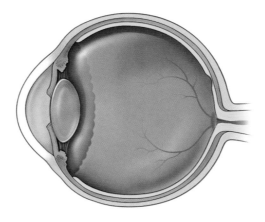

Figure 26.8 Major Features of the Human Eye

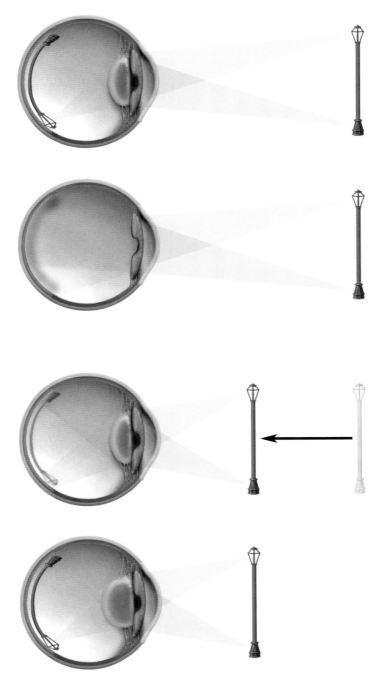

Figure 26.9 How the Human Eye Forms Sharp Images

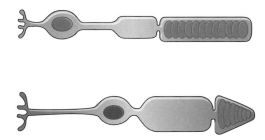

Figure 26.10 Human Photoreceptors

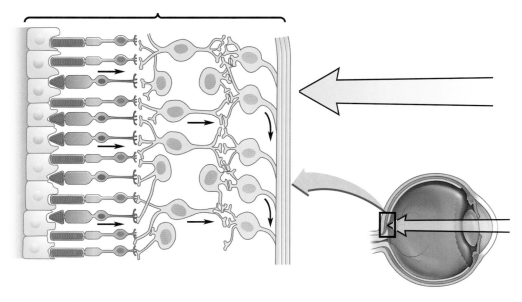

Figure 26.11 The Human Retina

WHAT IS A DRINK?
A standard drink is:

- One 12-ounce bottle of beer* or wine cooler
- One 5-ounce glass of wine
- 1.5 ounces of 80-proof distilled spirits

*Different beers have different alcohol content. Malt liquor has a higher alcohol content than most other brewed beverages.

CHAPTER 27 Muscles, Skeletons, and Movement

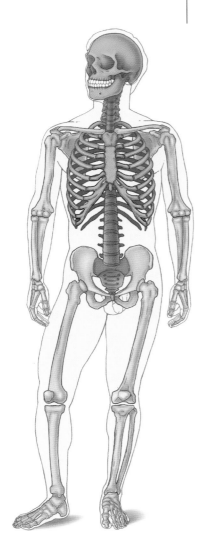

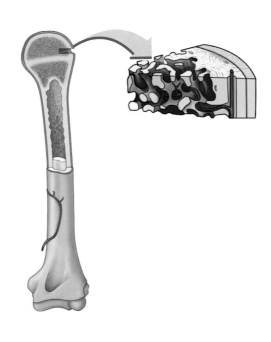

Figure 27.1 The Human Skeleton

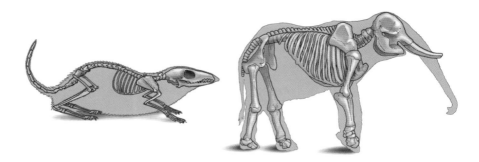

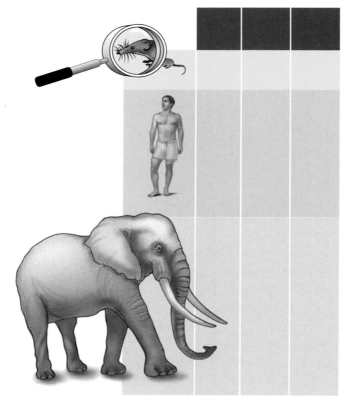

Figure 27.4 Animals Devote Different Proportions of Their Body Weight to Their Skeletons

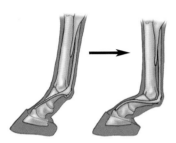

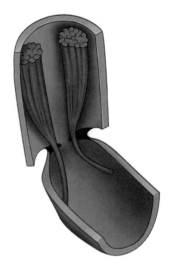

Figure 27.6 Endoskeletons and Exoskeletons

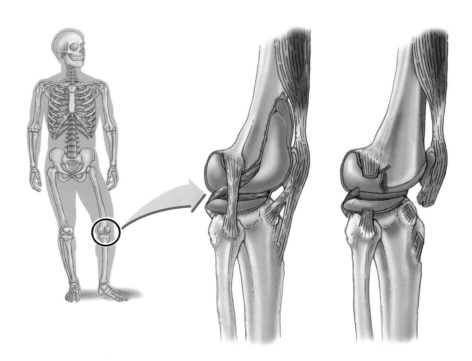

Figure 27.7 The Human Knee

Table 27.1

Common Ailments of the Knee Joint

Condition	Affects			
	Bone	Cartilage	Ligaments	Synovial sacs
Osteoarthritis	X	X		
Rheumatoid arthritis	X	X		X
Torn cartilage		X		
Torn ligament			X	
Bursitis				X

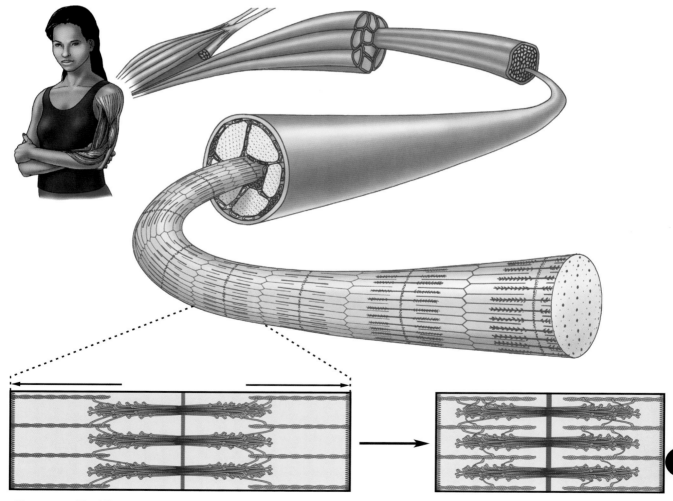

Figure 27.9 The Microscopic Structure of Muscle

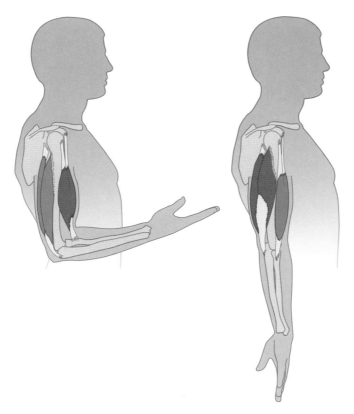

Figure 27.10 Muscles Work in Pairs

Table 27.2

Fast and Slow Muscle Fibers

	Type of muscle fiber	
Characteristic	Fast	Slow
Speed of contraction	Fast	Slow
Force of contraction	Weak	Powerful
Length of contraction	Brief	Sustained
Response of sarcomeres	Either no contraction or complete contraction	Partial contraction possible
Source of ATP	Fermentation	Aerobic respiration
Human example	The quadriceps muscle (in the thigh)	The gluteus maximus muscle (the largest muscle in the buttocks)

Figure 27.11 Mechanical and Biological Lever Systems

Figure 27.12 Adaptations of the Lever System in Mammalian Legs

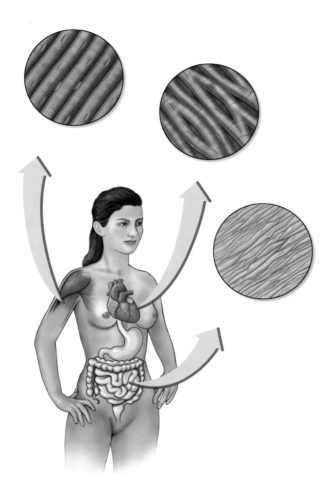

Figure 27.13 Some Specialized Muscle Types Create Internal Movements

Figure 27.15 Whales, Fish, and Birds Have Streamlined Designs

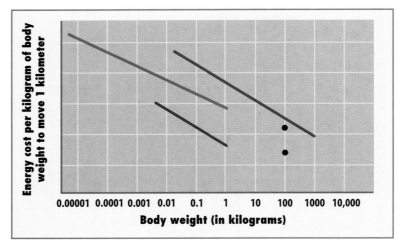

Figure 27.16 The Relative Costs of Locomotion

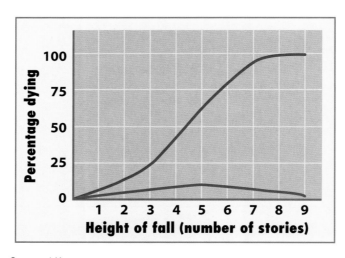

Figure 27.17 Survival of Falling Cats and Humans

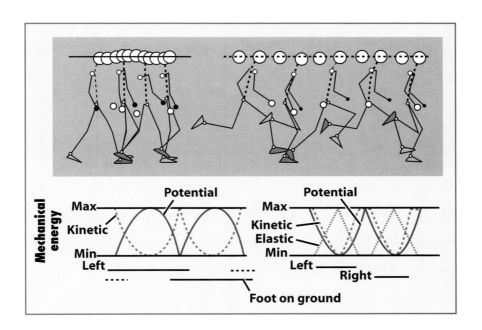

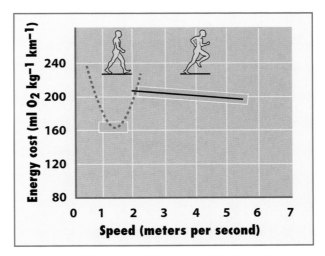

Science Toolkit Math Helps Us Understand the Energetics of Running and Walking in Humans

CHAPTER 28 | Defenses Against Disease

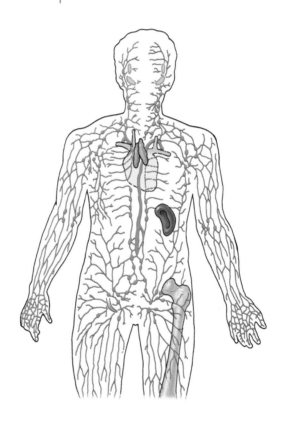

Figure 28.6 The Human Immune System

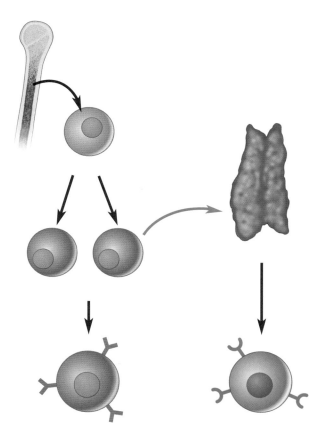

Figure 28.7 Lymphocyte Origins

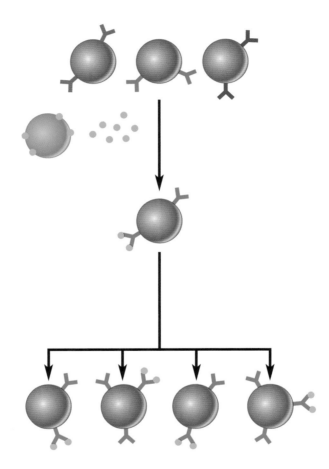

Figure 28.8 How the Human Body Recognizes Specific Invaders

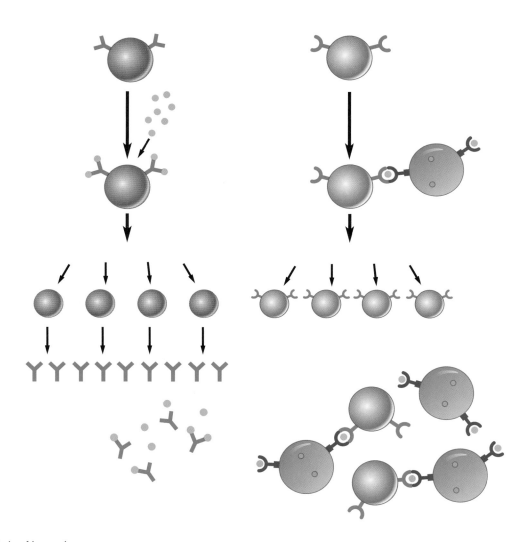

Figure 28.9 Two Kinds of Immunity

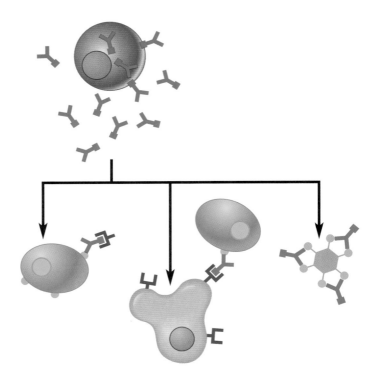

Figure 28.10 How Antibodies Work

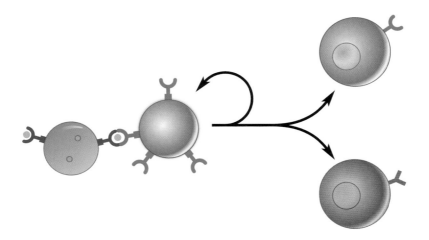

Figure 28.11 Helper T Cells

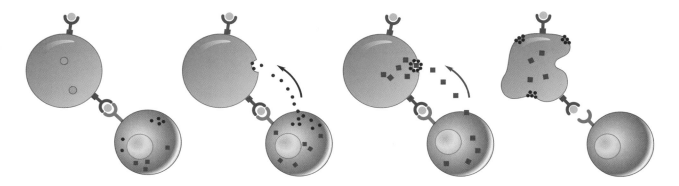

Figure 28.12 Killer T Cells Destroy Infected or Damaged Host Cells

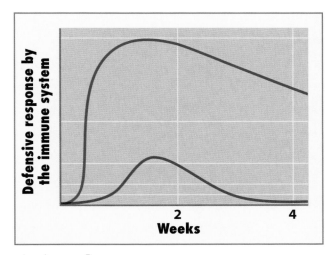

Figure 28.13 Primary versus Secondary Immune Responses

CHAPTER 29 Reproduction and Development

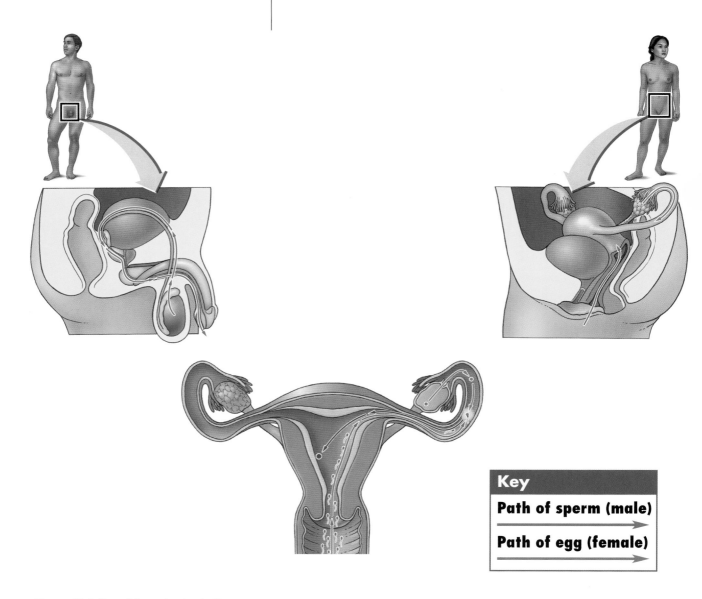

Key

Path of sperm (male) →

Path of egg (female) →

Figure 29.1 Sexual Reproduction in Humans

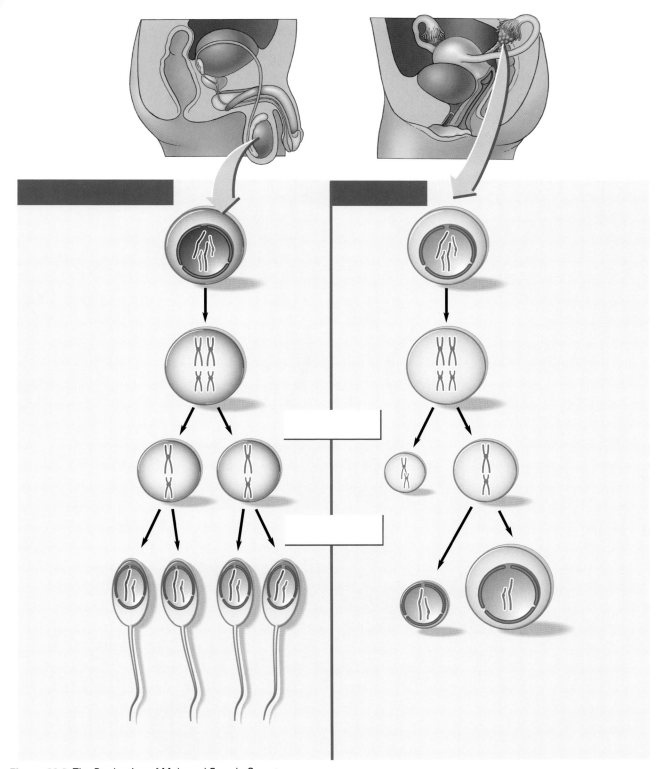

Figure 29.2 The Production of Male and Female Gametes

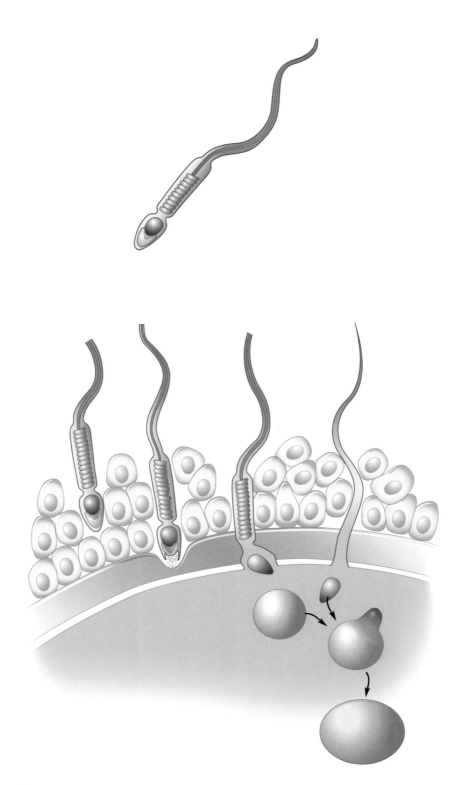

Figure 29.3 Fertilization Occurs When a Single Sperm Enters the Egg

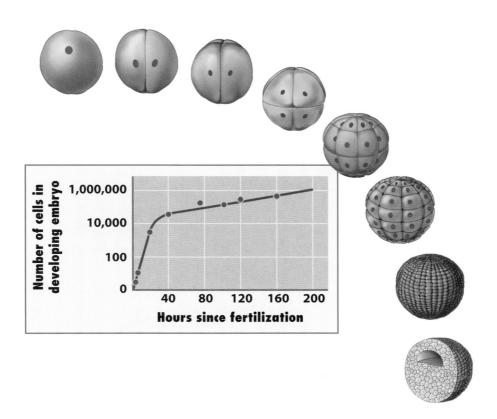

Figure 29.8 Early Development of Vertebrates

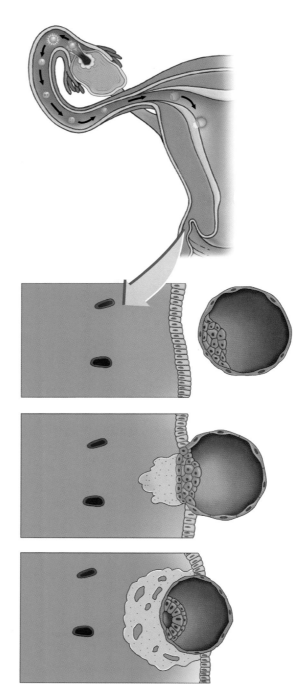

Figure 29.9 Implantation in Mammalian Development

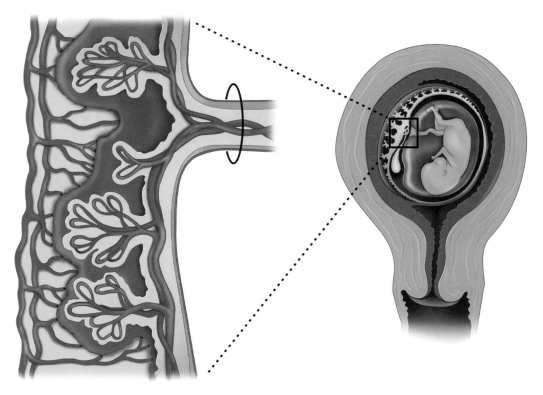

Figure 29.10 The Placenta

Table 29.1

The Fates of the Three Cell Layers in the Developing Embryo

Cell layer	Corresponding adult structures
Endoderm	Liver, pancreas, thyroid, and linings of the gut and lungs
Mesoderm	Skeleton, muscles, reproductive structures, kidneys, circulatory and lymphatic systems, blood, and the inner layer of skin
Ectoderm	Skull, nerves and brain, the outer layer of skin, and teeth

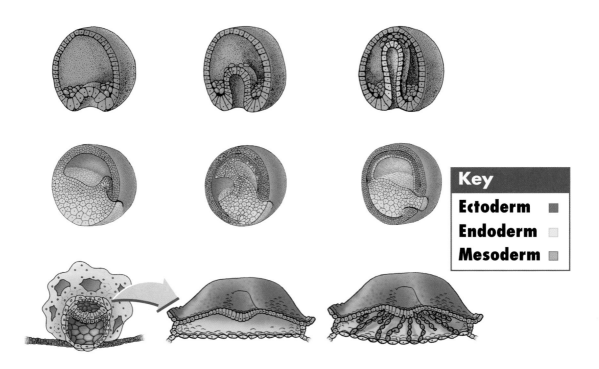

Figure 29.11 The Three Cell Layers Have Developmentally Distinct Fates

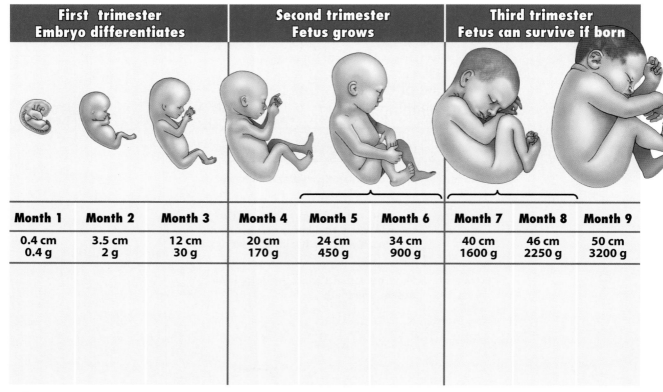

First trimester Embryo differentiates			Second trimester Fetus grows			Third trimester Fetus can survive if born		
Month 1	**Month 2**	**Month 3**	**Month 4**	**Month 5**	**Month 6**	**Month 7**	**Month 8**	**Month 9**
0.4 cm 0.4 g	3.5 cm 2 g	12 cm 30 g	20 cm 170 g	24 cm 450 g	34 cm 900 g	40 cm 1600 g	46 cm 2250 g	50 cm 3200 g

Figure 29.12 Development of the Human Embryo and Fetus

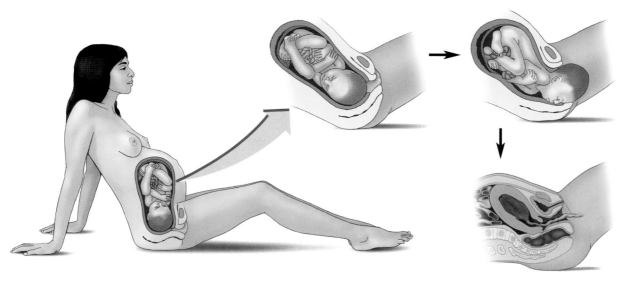

Figure 29.13 The Birth Process

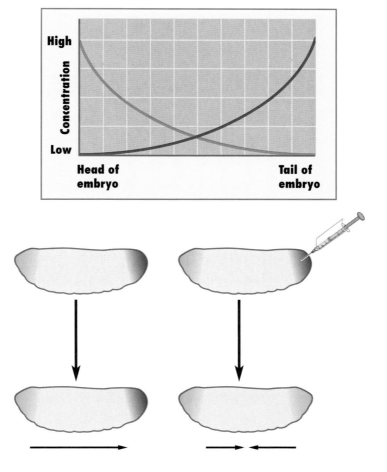

Figure 29.14 Morphogens Often Control Development

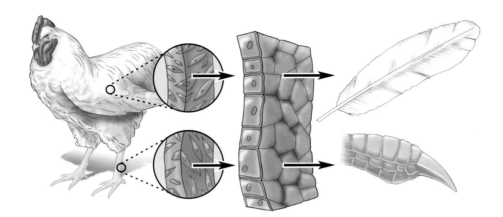

Figure 29.15 The Identity of Surrounding Cells Influences Development

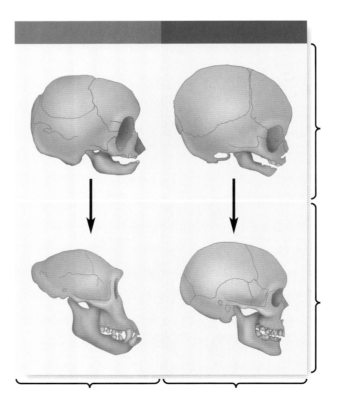

Figure 29.17 Evolution of the Human Skull

Table 29.2

The Effect of Age on the Reproductive Success of Females

Age (years)	Cycles that result in a pregnancy (%)	Cycles that result in an embryo that fails to complete development (%)
Under 30	29.0	14.9
30–35	19.8	16.5
35–40	17.1	22.4
40–45	12.8	33.2

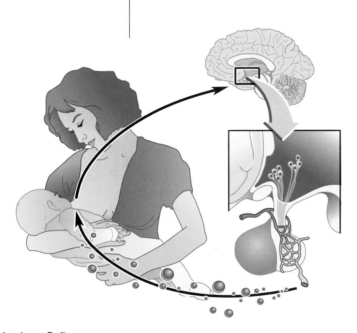

Figure 30.1 The Milk-Letdown Reflex

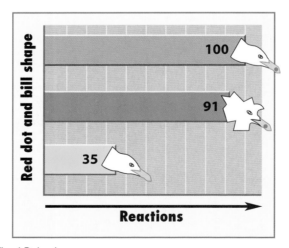

Figure 30.2 Simple Stimuli Trigger Fixed Behaviors

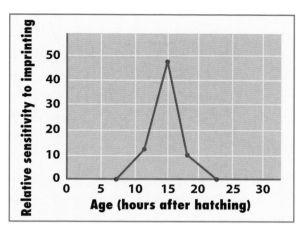

Figure 30.3 Some Animals Learn Who Their Parents Are by Imprinting

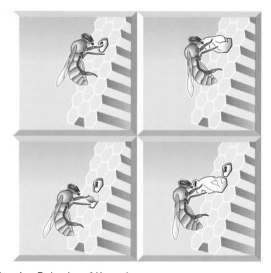

Figure 30.4 Genes Control the Nest-Cleaning Behavior of Honeybees

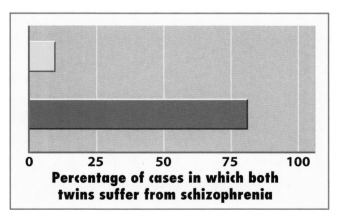

Figure 30.5 The Genetics of Schizophrenia

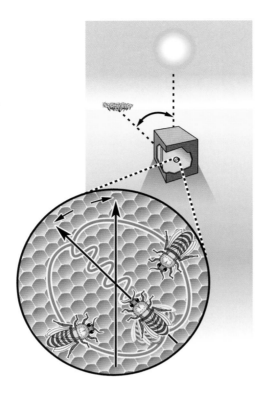

Figure 30.7 Honeybees Communicate by Dancing in Their Hive

INTERLUDE **E** Smoking: Beyond Lung Cancer

Table E.1

How Tobacco Smoke Kills

Illness	Annual deaths worldwide due to tobacco smoke
Diseases of the circulatory system	1,690,000
Cancers	1,469,000
Respiratory diseases	1,399,000

Table E.2

The Major Components of Cigarette Smoke with Adverse Health Effects

Component	Known effects
Smoke particulates	Increase blood clotting, irritate respiratory system
Carbon monoxide	Reduces oxygen-carrying capacity of blood
Polycyclic aromatic hydrocarbons	Carcinogenic
Aza-arenes [AZZ-uh-AH-reens]	Carcinogenic
N-Nitrosamines [EN-nigh-TROZ-uh-meens]	Carcinogenic
Aromatic amines	Carcinogenic
Heterocyclic aromatic amines	Carcinogenic
Aldehydes	Carcinogenic
Inorganic compounds (such as arsenic)	Carcinogenic
Miscellaneous tar constituents	Stain skin and teeth; may have other effects
Cyanide	Toxic
Formaldehyde	Toxic
Ammonia	Toxic
Nicotine	Toxic; addictive drug

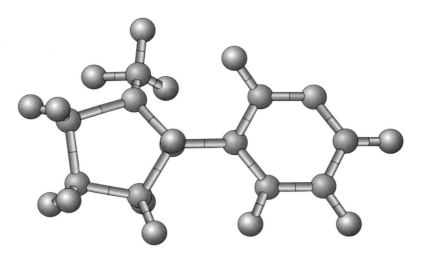

Figure E.7 The Chemical Structure of Nicotine

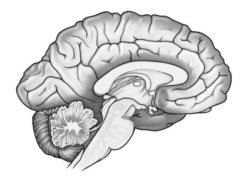

Figure E.8 Addictive Drugs Stimulate Cells in the Nucleus Accumbens

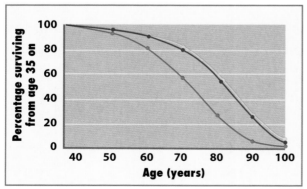
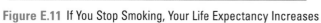
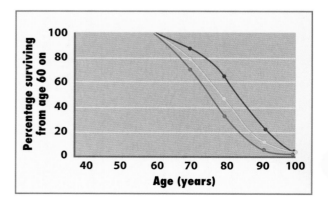

Figure E.11 If You Stop Smoking, Your Life Expectancy Increases

CHAPTER 31

Plant Structure, Nutrition, and Transport

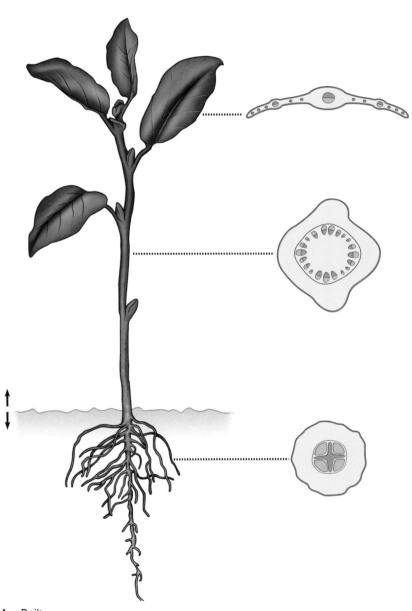

Figure 31.2 How Plants Are Built

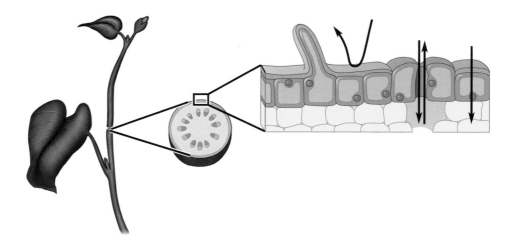

Figure 31.3 Dermal Tissue Is Where Plants Meet the Environment

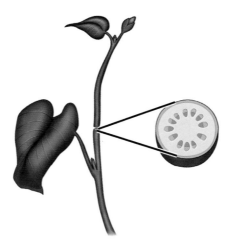

Figure 31.4 Ground Tissue Makes Up the Bulk of the Plant Body

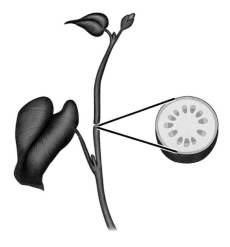

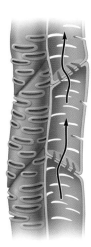

Figure 31.5 Vascular Tissue Transports Food and Water

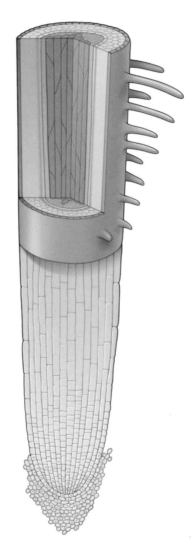

Figure 31.6 Anchoring the Plant and Absorbing Nutrients

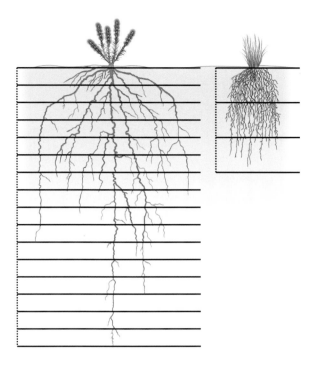

Figure 31.7 Two Kinds of Root Systems

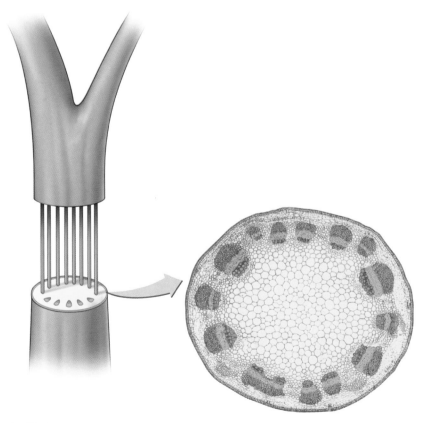

Figure 31.8 Stems Support the Plant

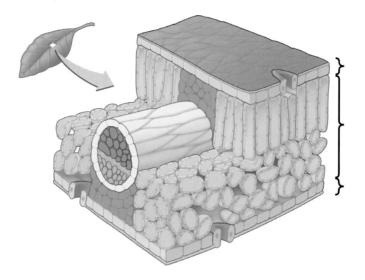

Figure 31.9 Leaves Feed the Plant

Figure 31.10 How Plants Get Raw Materials for Growth

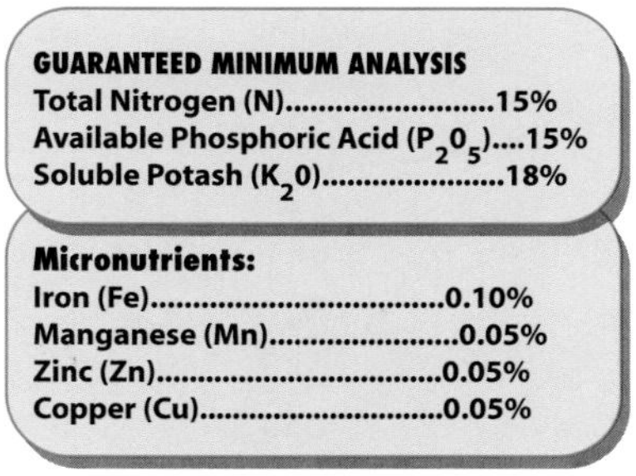

GUARANTEED MINIMUM ANALYSIS
Total Nitrogen (N)..........................15%
Available Phosphoric Acid (P_2O_5)....15%
Soluble Potash (K_2O).......................18%

Micronutrients:
Iron (Fe)....................................0.10%
Manganese (Mn)......................0.05%
Zinc (Zn)...................................0.05%
Copper (Cu)..............................0.05%

Figure 31.11 What Plants Eat

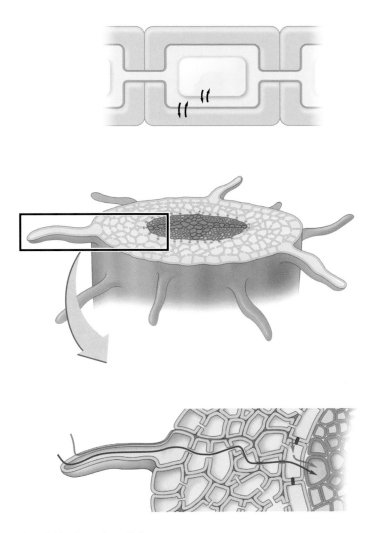

Figure 31.12 How Plants Absorb Nutrients from Soil

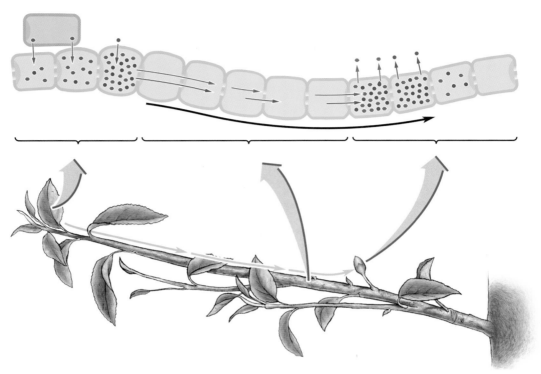

Figure 31.14 How Food Is Transported in Phloem Tissue

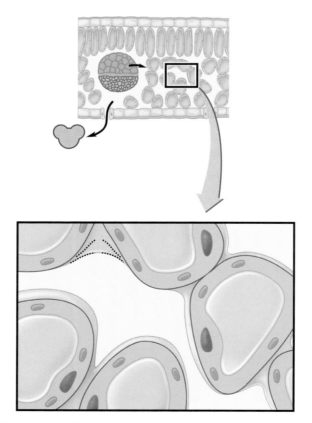

Figure 31.15 How Plants Move Water in Xylem Tissue

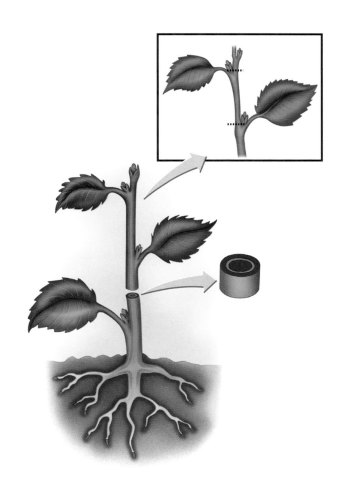

Figure 32.1 How Plants Grow

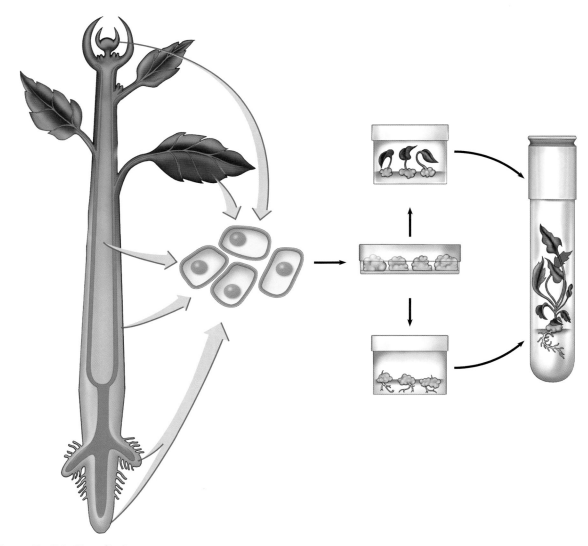

Science Toolkit Plant Cloning

Figure 32.2 Growing Up and Out

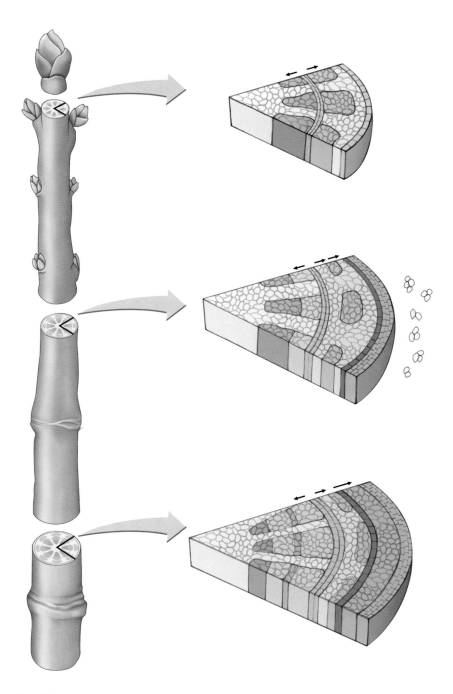

Figure 32.3 Growing Thicker Over Time

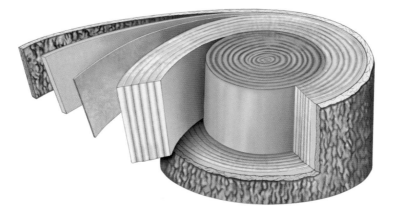

Figure 32.4 The Insides of a Tree

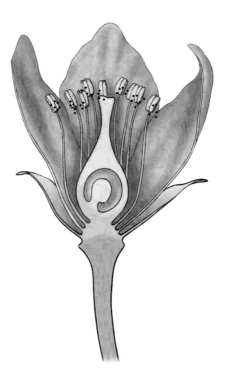

Figure 32.5 Four Whorls Make a Flower

Figure 32.6 From Generation to Generation: The Life Cycle of a Flowering Plant

Figure 32.7 Bribing Animals to Do the Work

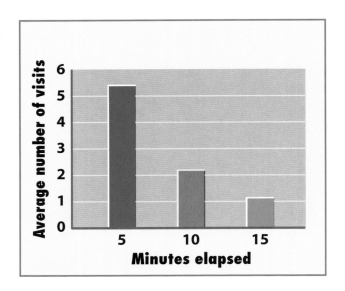

Figure 32.8 Sexual Trickery

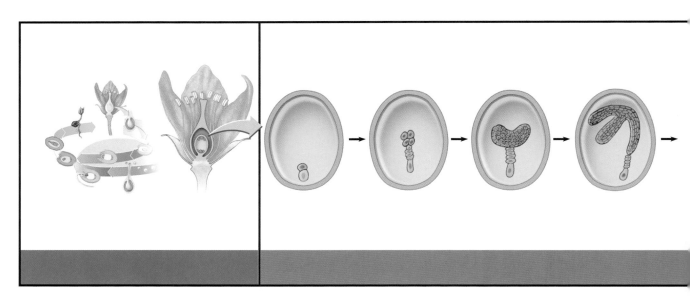

Figure 32.9 From Zygote to Seedling

Figure 32.11 Removing the Tip Lets the Branches Grow

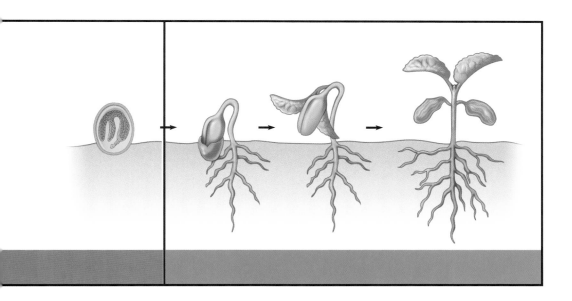

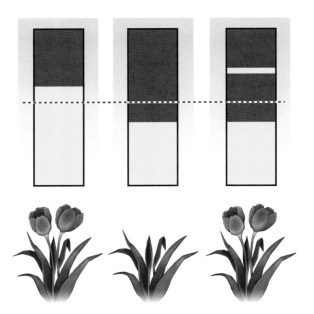

Figure 32.12 Plants Can Detect Day Length

Figure 32.13 Hormones Control Flowering

Feeding a Hungry Planet

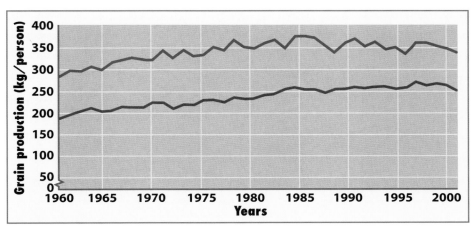

Figure F.3 A Drop in World Grain Production

Table F.1

Food Production from the Three Traditional Food Systems

In the second half of the twentieth century, world food production per person from each of the three traditional food systems grew rapidly, then declined.

Food system	Food	Years of growth	Growth per person (%)	Years of decline	Decline from peak (%)
Cropland	Grain	1950–1984	+38	1984–2000	−11
Rangeland	Beef and mutton	1950–1972	+44	1972–2000	−15
Oceanic fisheries	Seafood	1950–1988	+112	1988–1998	−17

Table F.2

Sources of Animal Protein

"Annual rate of growth" is the average yearly increase (1990–2000) in world production of animal protein from five sources: aquaculture, three feedlot systems (poultry, pork, and beef), and oceanic fisheries. In aquaculture and the three feedlot systems, increases in production were greatest for those sources that use less grain per kilogram (kg) of animal live weight produced.

Source of animal protein	Annual rate of growth (%)	Amount of grain (in kg) needed to produce 1 kg animal live weight
Aquaculture	11.4	1.5–2
Poultry	4.9	2
Pork	2.5	4
Beef	0.5	7
Oceanic fisheries	0.1	Not applicable

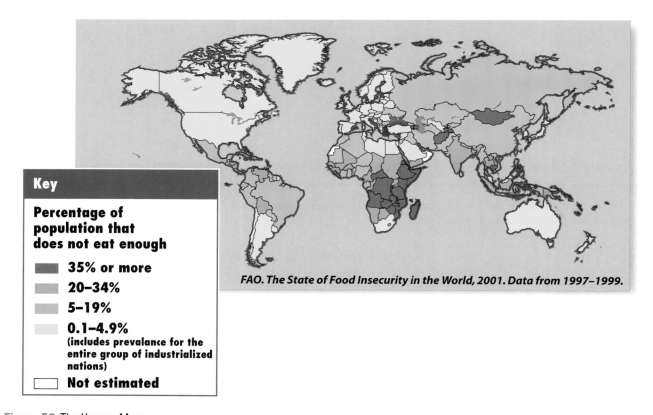

Key

Percentage of population that does not eat enough

- 35% or more
- 20–34%
- 5–19%
- 0.1–4.9% (includes prevalance for the entire group of industrialized nations)
- Not estimated

FAO. The State of Food Insecurity in the World, 2001. Data from 1997–1999.

Figure F.5 The Hunger Map

Selected whole food sources of vitamin E	Mg
Almonds, 1 oz. (24 nuts)	7.4
Hazelnuts, 1 oz. (20 nuts)	4.3
Canola oil, 1 tbsp.	2.9
Broccoli, 1 cup cooked	2.6
Peanuts, 1 oz. (28 nuts)	2.2
Olive oil, 1 tbsp.	1.7
Wheat germ, 1 tbsp.	1.3
Red bell pepper, 1 cup	1.0
Kiwifruit, 1 medium	0.9
Olives, 5 large	0.7
Spinach, 1 cup raw	0.6
Avocado, 1 oz.	0.4
Brown rice, 1 cup, long-grain	0.4
Apple, 1 medium	0.4
Banana, 1 medium	0.3
Sesame seeds, 1 tbsp.	0.2
Romaine lettuce, 1 cup	0.2
Beef, ground, 3 oz.	0.2

Biology Matters The Importance of Vitamin E

Figure F.11 How Farm Ecosystems Work

CHAPTER 33

The Biosphere

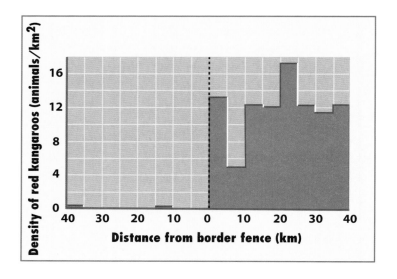

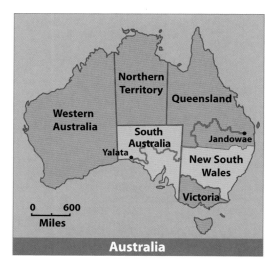

Figure 33.2 An Explosion in Numbers

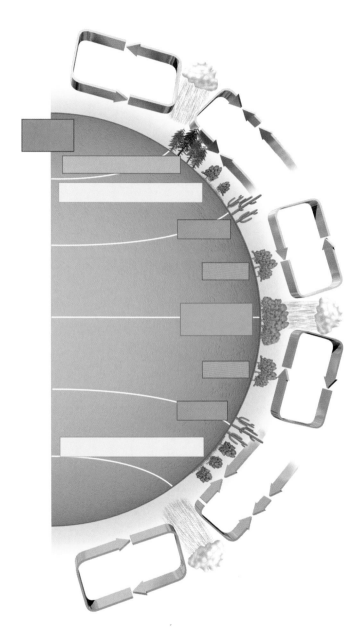

Figure 33.3 Earth Has Four Giant Convection Cells

Figure 33.4 Global Patterns of Air Circulation

Figure 33.5 The World's Major Ocean Currents

Figure 33.6 The Rain Shadow Effect

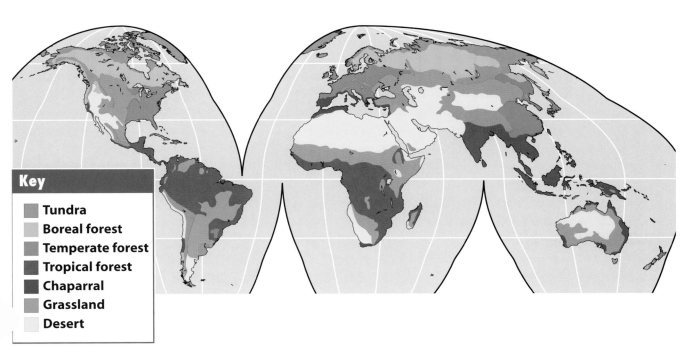

Key
- Tundra
- Boreal forest
- Temperate forest
- Tropical forest
- Chaparral
- Grassland
- Desert

Figure 33.7 Major Terrestrial Biomes

Figure 33.10 The Location of Terrestrial Biomes Depends on Temperature and Precipitation

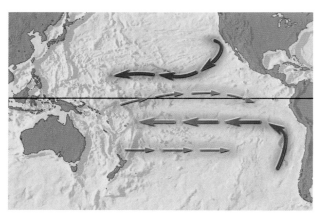
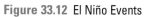
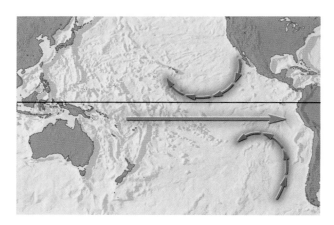

Figure 33.12 El Niño Events

34 Growth of Populations

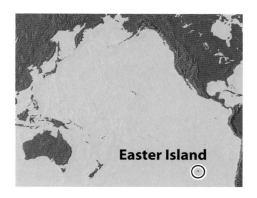

Chapter 34 Opener Easter Island

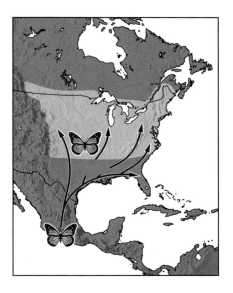

Figure 34.2 Before the Crash

Species	Annual Proportional Increase (λ)	Time for Population to Grow from 2 to More Than 1,000,000,000 Individuals	Time Required to Produce More Individuals Than There Are Atoms in the Universe
Humans (in 2003)	1.012	1,680 years	15,385 years
Wild ginger	1.1	211 years	1,926 years
Reindeer	1.2	110 years	1,007 years
Gray kangaroos	1.9	32 years	286 years
Field voles	24	7 years	58 years
Rice weevils	10^{17}	7 months	5 years
E. coli bacteria	10^{5274}	15 hours	6 days

Biology Matters How Fast Can Populations Grow?

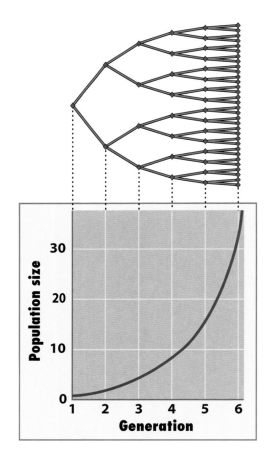

Figure 34.3 Exponential Growth

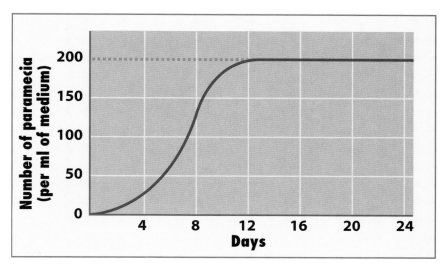

Figure 34.6 Carrying Capacity

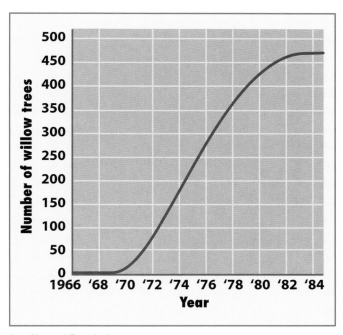

Figure 34.7 An S-Shaped Curve in a Natural Population

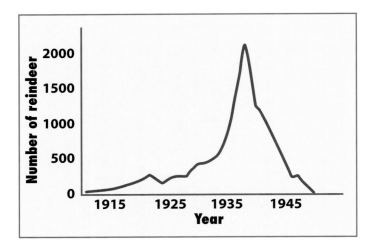

Figure 34.8 Boom and Bust

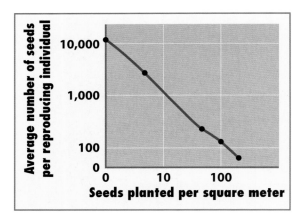

Figure 34.9 It's Getting Crowded

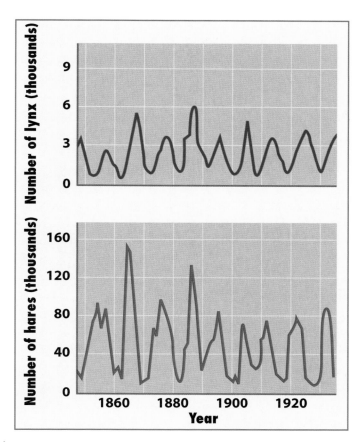

Figure 34.10 Population Cycles

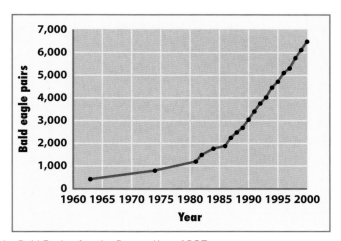

Science Toolkit Recovery of the Bald Eagle after the Ban on Use of DDT

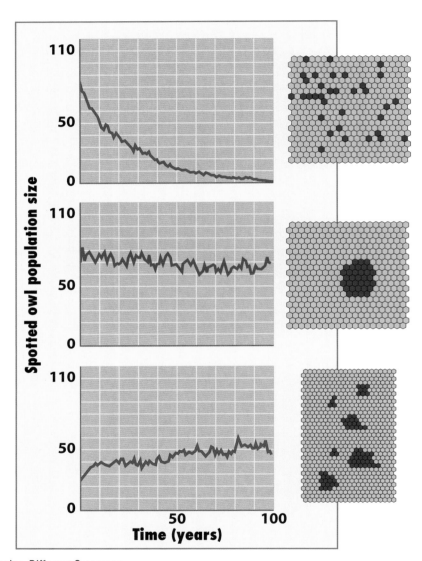

Figure 34.11 Same Species, Different Outcomes

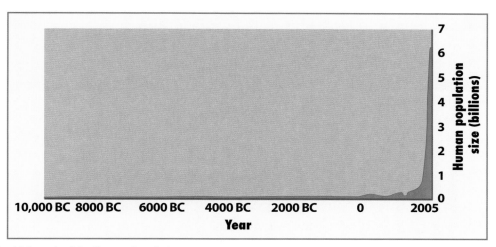

Figure 34.12 Rapid Growth of the Human Population

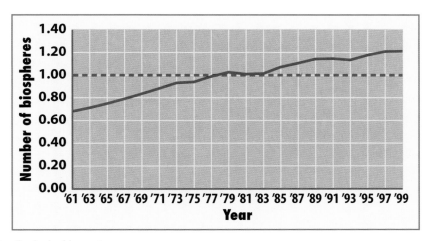

Figure 34.13 Our Rising Ecological Impact

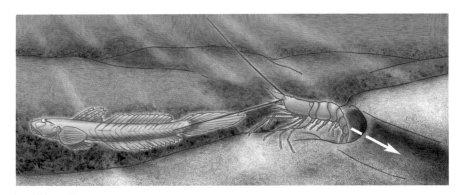

Figure 35.1 A Behavioral Mutualism

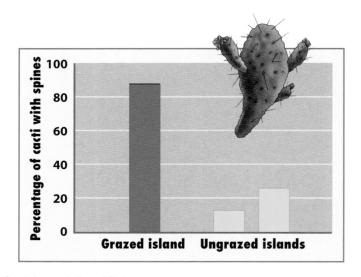

Figure 35.5 Spines on Some Cacti Are an Induced Defense

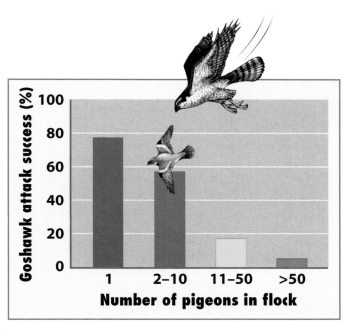

Figure 35.8 Safety in Numbers

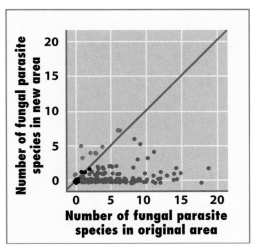

Figure 35.9 Leaving Their Parasites Behind

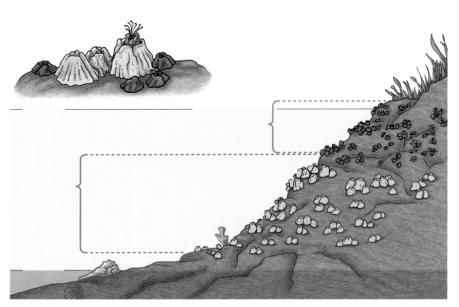

Figure 35.10 What Keeps Them Apart?

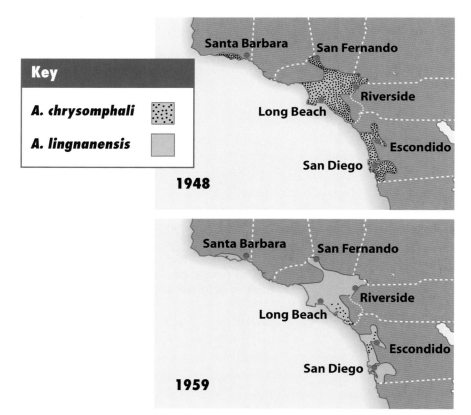

Figure 35.11 A Superior Competitor Moves In

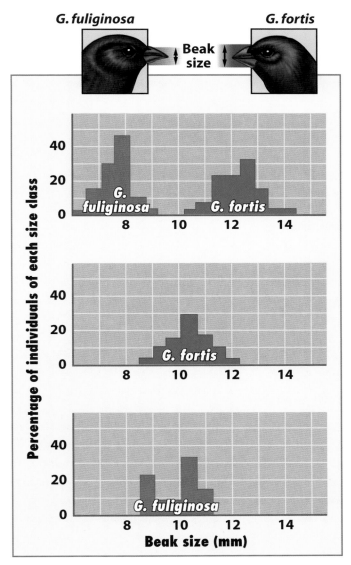

Figure 35.12 Character Displacement

Chapter 36 Opener A Natural Experiment

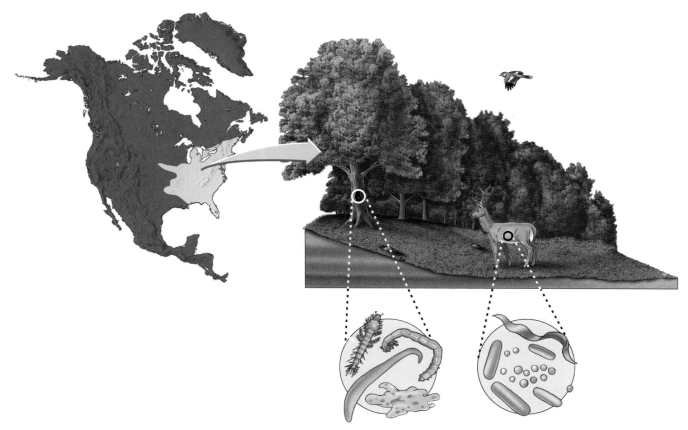

Figure 36.1 Ecological Communities

Figure 36.2 Which Community Has Greater Diversity?

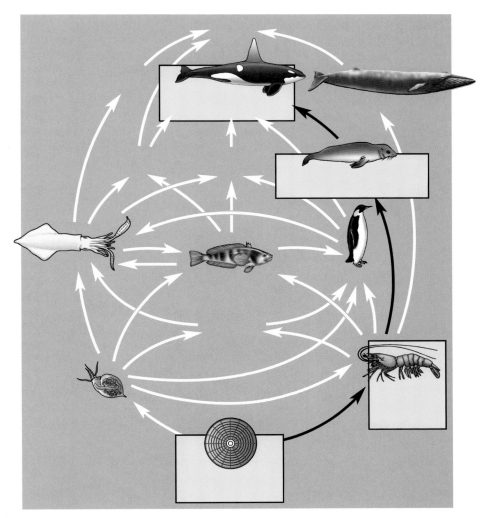

Figure 36.3 A Food Web

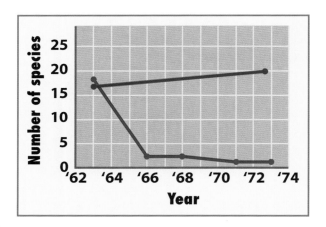

Figure 36.4 A Keystone Species

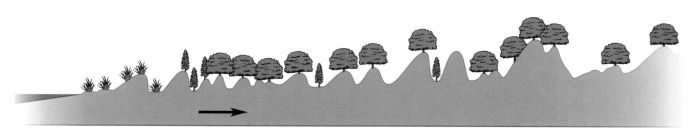

Figure 36.5 Succession

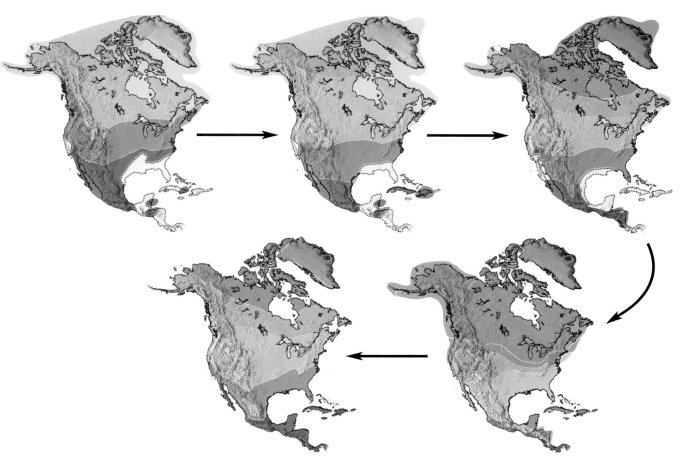

Figure 36.7 Climates Change, Communities Change

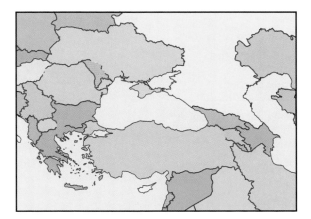

Figure 36.9 Changes in the Black Sea

CHAPTER 37 Ecosystems

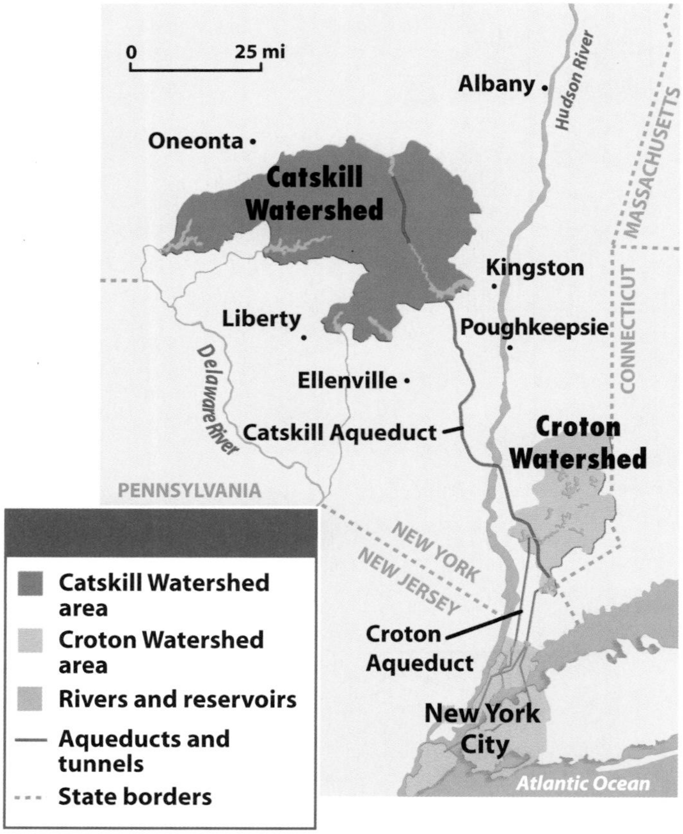

0
25 mi
Albany
Hudson River
Oneonta
Catskill Watershed
MASSACHUSETTS
Kingston
Liberty
Poughkeepsie
CONNECTICUT
Delaware River
Ellenville
Catskill Aqueduct
Croton Watershed
PENNSYLVANIA
NEW YORK
NEW JERSEY
Catskill Watershed area
Croton Watershed area
Rivers and reservoirs
Aqueducts and tunnels
State borders
Croton Aqueduct
New York City
Atlantic Ocean

Chapter 37 Opener New York City's Water Supply System

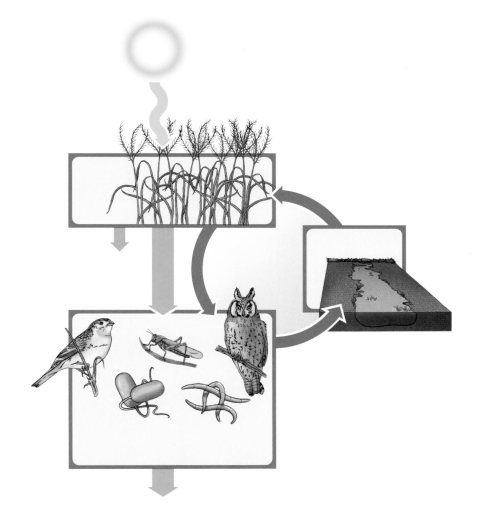

Figure 37.1 How Ecosystems Work

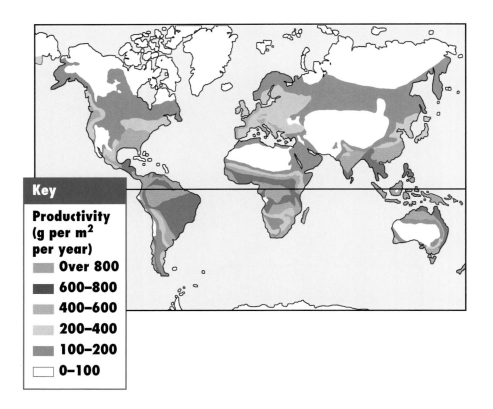

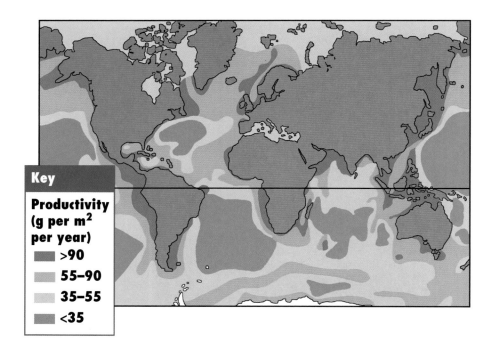

Figure 37.3 Net Primary Productivity Varies Across the Globe

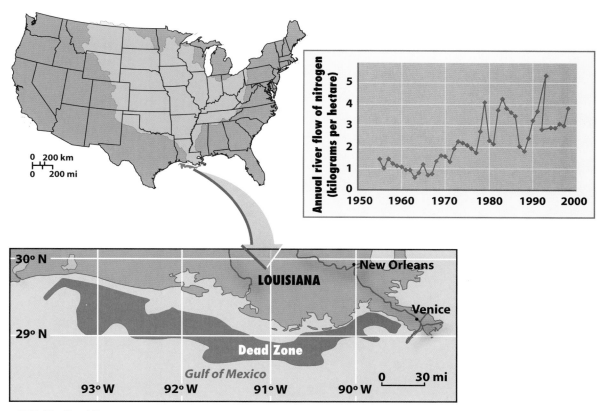

Figure 37.5 The Dead Zone

Figure 37.6 An Idealized Energy Pyramid

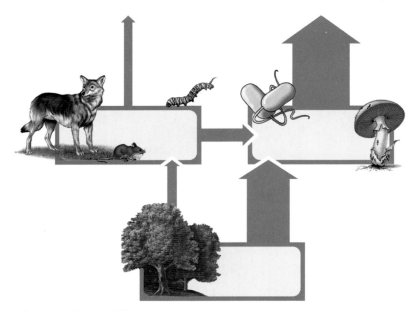

Figure 37.7 Decomposers Consume Most of NPP

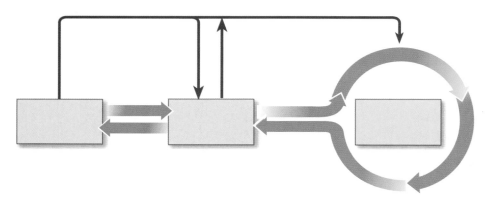

Figure 37.8 Nutrient Cycling

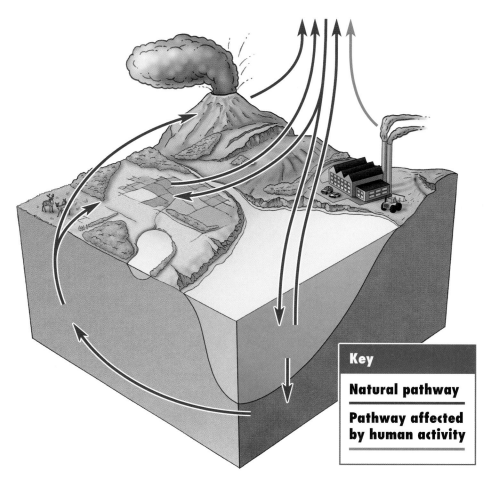

Figure 37.9 The Sulfur Cycle

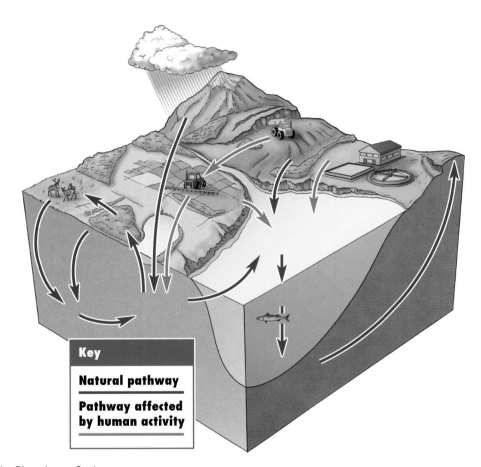

Key

Natural pathway

**Pathway affected
by human activity**

Figure 37.10 The Phosphorus Cycle

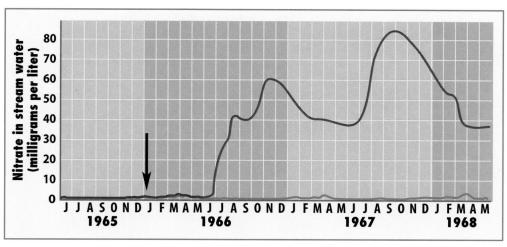

Figure 37.11 Altering Nutrient Cycles in a Forest Ecosystem

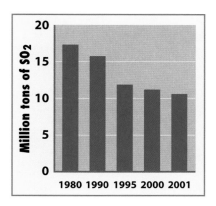

Figure 37.12 Acid Rain

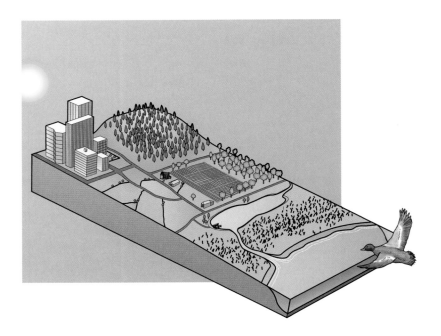

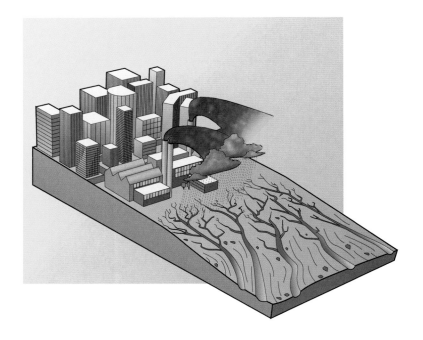

Figure 37.15 Services Provided by Ecosystems

Biome	Global Area (hectares × 10^6)	Ecosystem Services (1994 US $ per hectare, per year)					Total Global Value (billions per year)
		Food Production	Water Regulation	Water Supply	Waste Treatment	Recreation	
Lakes/ rivers	200	41	5,445	2,117	665	230	1,700

Biology Matters How Much Are Lakes and Rivers Worth?

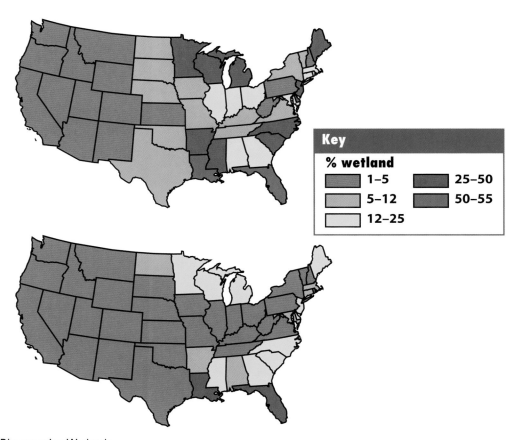

Figure 38.2 Disappearing Wetlands

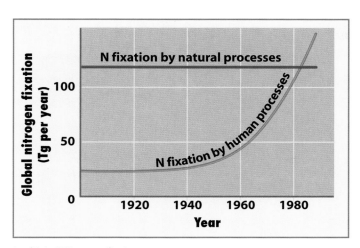

Figure 38.4 Human Effects on the Global Nitrogen Cycle

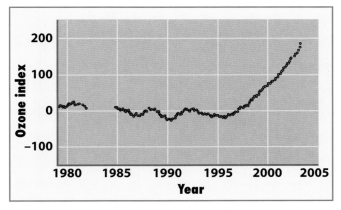

Science Toolkit The Ozone Layer Begins to Recover

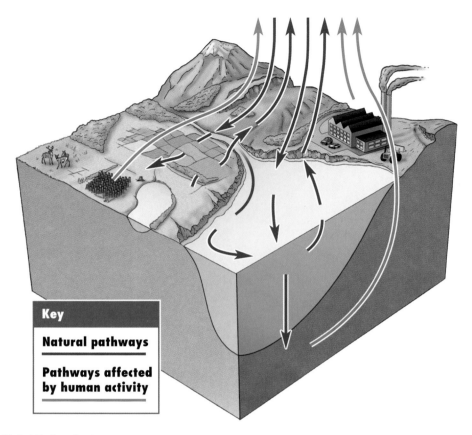

Figure 38.6 The Global Carbon Cycle

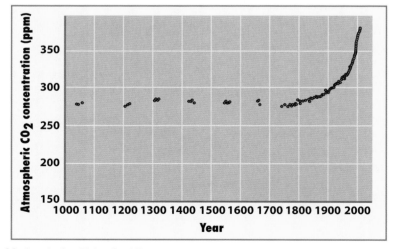

Figure 38.7 Atmospheric CO_2 Levels Are Rising Rapidly

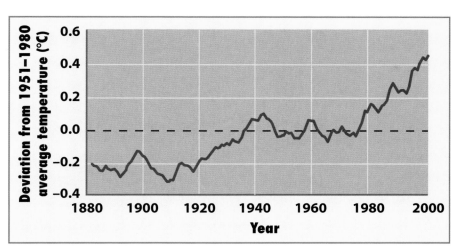

Figure 38.9 Global Temperatures Are on the Rise

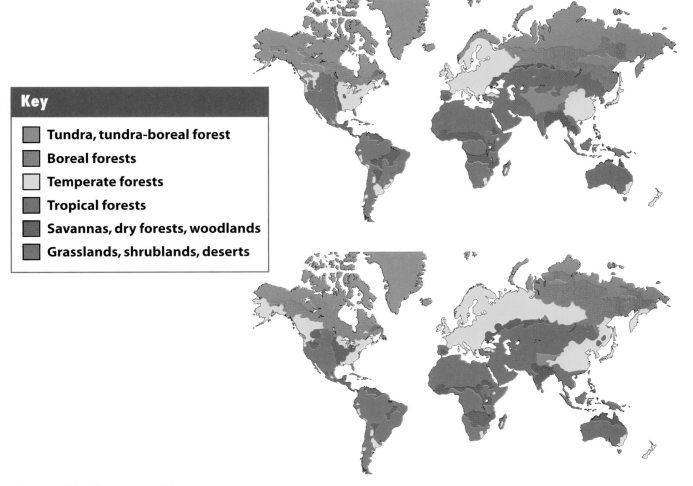

Key

- Tundra, tundra-boreal forest
- Boreal forests
- Temperate forests
- Tropical forests
- Savannas, dry forests, woodlands
- Grasslands, shrublands, deserts

Figure 38.10 Biomes on the Move

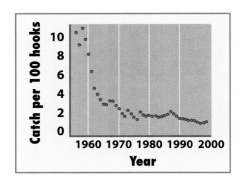

Figure 38.11 Declining Fish Populations

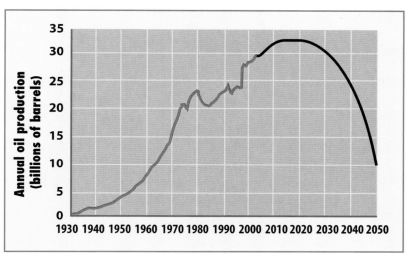

Figure G.2 Running Out of Oil

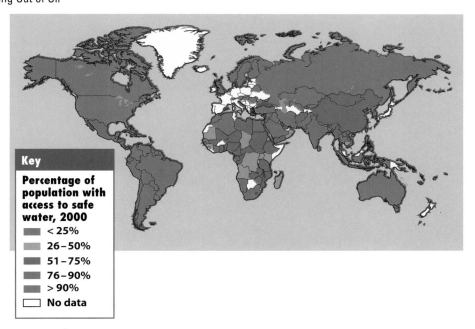

Key

Percentage of population with access to safe water, 2000

- < 25%
- 26–50%
- 51–75%
- 76–90%
- > 90%
- No data

Figure G.3 Water Quality Varies Across the Globe

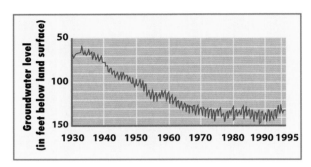

Figure G.4 Declining Groundwater Levels

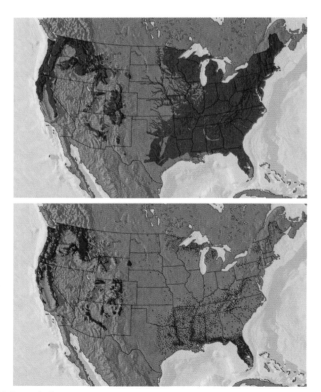

Figure G.5 The Destruction of Old-Growth Forests

Figure G.7 Greenroofs Around the World

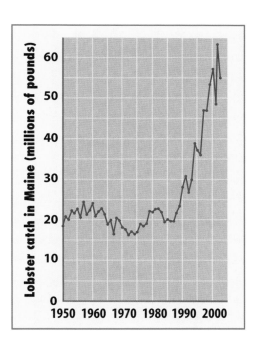

Figure G.12 A Sustainable Fishery

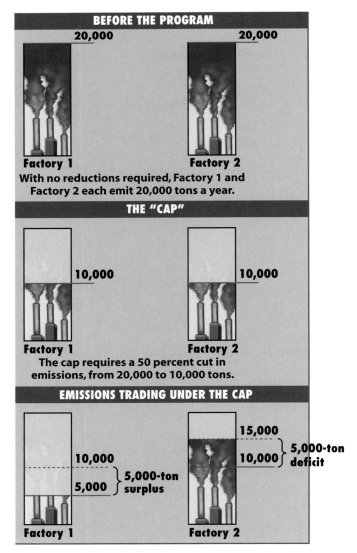

Figure G.13 How a Cap-and-Trade System Works

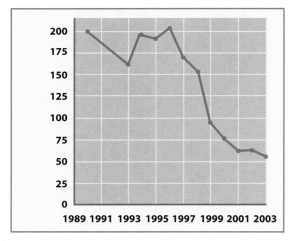

Figure G.14 Declining Greenhouse Gas Emissions